Methods in Molecular Biology

Series Editor
John M. Walker
School of Life and Medical Sciences
University of Hertfordshire
Hatfield, Hertfordshire, AL10 9AB, UK

For further volumes:
http://www.springer.com/series/7651

Vibrio Cholerae

Methods and Protocols

Edited by

Aleksandra E. Sikora

Department of Pharmaceutical Sciences, College of Pharmacy, Oregon State University, Corvallis, OR, USA

Humana Press

Editor
Aleksandra E. Sikora
Department of Pharmaceutical Sciences
College of Pharmacy
Oregon State University
Corvallis, OR, USA

ISSN 1064-3745 ISSN 1940-6029 (electronic)
Methods in Molecular Biology
ISBN 978-1-4939-8684-2 ISBN 978-1-4939-8685-9 (eBook)
https://doi.org/10.1007/978-1-4939-8685-9

Library of Congress Control Number: 2018948674

Cover illustration: This is V. cholerae N16961. Transmission electron microscopy photo by Suzanne. R. Lybarger and Aleksandra E. Sikora.

Printed on acid-free paper

This Humana Press imprint is published by the registered company Springer Science+Business Media, LLC part of Springer Nature.
The registered company address is: 233 Spring Street, New York, NY 10013, U.S.A.

Preface

Vibrio cholerae is a member of a range of natural aquatic ecosystems and serves as the most prominent of a number of *Vibrio* species. *V. cholerae* causes the devastating diarrheal disease, cholera, primarily in the developing countries of Asia, Africa, and other regions disrupted by climate change, other natural calamity, or war. The persistence of *V. cholerae* in natural habitats is a crucial factor in the epidemiology of cholera. Cholera remains one of the top global infectious disease threats and, therefore, understanding the fundamental mechanisms of *V. cholerae* virulence and environmental survival is a prerequisite for the development of new therapeutic treatments or preventative measures.

This book was prompted by the significance of *V. cholerae* as an infectious agent and the emergence of new methodologies. The scope of this book covers various approaches applied to studying the biology of *V. cholerae*: from basic microbiological techniques including laboratory maintenance, culturing, and phenotypic assays, different animal models (zebrafish, infant mice, fruit fly) developed to elucidate pathogenesis and determinants of environmental survival, isolation and characterization of outer membrane vesicles, to holistic approaches incorporating transcriptomics, proteomics, metabolomics, and high-throughput screening of natural product libraries. Thus, the contents of this book offer a head start for the novice microbiologist with the focus on *V. cholerae* as well as an enhancement of skills for experienced researchers and teachers of microbiology. The book is organized into independent and self-contained chapters, with an introduction that describes the methodology, clearly outlined materials and procedures, and notes to help the reader with potential troubleshooting.

I would like to thank, first and foremost, the expert friends and wise colleagues who shared their knowledge, time, and patience and contributed the chapters to this book. I want to acknowledge Prof. John Walker for his encouragement and guidance throughout the editing process as well as Ms. Anna Rakovsky and Kim Gratz for their excellent assistance in assembling the book. Lastly, but not least importantly, I want to thank my husband, Tim Weber, whose continuous love, encouragement, support, and vision stimulated and inspired me through the creative process of this book.

Corvallis, OR, USA ***Aleksandra E. Sikora***

Contents

Preface ..... *v*
Contributors ..... *ix*

1 Laboratory Culturing Techniques and Maintenance of *Vibrio cholerae* 1
Kyle D. Brumfield, Bailey M. Carignan, and Mike S. Son

2 Genotypic and Phenotypic Assays to Distinguish *Vibrio cholerae* Biotype 11
Kyle D. Brumfield, Bailey M. Carignan, and Mike S. Son

3 Preparation of *Vibrio cholerae* Samples for RNA-seq Analysis 29
Jyl S. Matson

4 Random Transposon Mutagenesis of *Vibrio cholerae* 39
Jyl S. Matson

5 Metabolomics of *Vibrio cholerae* 45
Yusuke Minato, Jay S. Kirkwood, and Claudia C. Häse

6 Natural Cotransformation and Multiplex Genome Editing by Natural Transformation (MuGENT) of *Vibrio cholerae* 53
Ankur B. Dalia

7 Chromatin Immunoprecipitation 65
Julio C. Ayala, Jorge A. Benitez, and Anisia J. Silva

8 Fly Models of *Vibrio cholerae* Infection and Colonization 77
Alexandra E. Purdy

9 *Danio rerio* as a Native Host Model for Understanding Pathophysiology of *Vibrio cholerae* 97
Kristie C. Mitchell and Jeffrey H. Withey

10 Transposon Sequencing of *Vibrio cholerae* in the Infant Rabbit Model of Cholera 103
Lauren M. Shull and Andrew Camilli

11 Isolation of Outer Membrane Vesicles Including Their Quantitative and Qualitative Analyses 117
Paul Kohl, Franz G. Zingl, Thomas O. Eichmann, and Stefan Schild

12 Utilization of *Vibrio cholerae* as a Model Organism to Screen Natural Product Libraries for Identification of New Antibiotics 135
Aleksandra E. Sikora, Richard Tehan, and Kerry McPhail

13 Infant Mouse Model of *Vibrio cholerae* Infection and Colonization 147
Jyl S. Matson

14 Methods for Assessments of Collagenolytic Activity of the *Vibrio cholerae* Extracellular Proteases, Purification of Secreted Collagenase VchC, and Extraction of Type I Collagen from Fish Skin 153
Bo R. Weber and Aleksandra E. Sikora

15 Proteomics of *Vibrio cholerae* 171
Ryszard A. Zielke

Index *179*

Contributors

JULIO C. AYALA • *Department of Microbiology, Biochemistry and Immunology, Morehouse School of Medicine, Atlanta, GA, USA; Department of Microbiology, University of Alabama at Birmingham, Birmingham, AL, USA*

JORGE A. BENITEZ • *Department of Microbiology, Biochemistry and Immunology, Morehouse School of Medicine, Atlanta, GA, USA*

KYLE D. BRUMFIELD • *Department of Biological Sciences, Plymouth State University, Plymouth, NH, USA*

ANDREW CAMILLI • *Tufts University School of Medicine, Boston, MA, USA*

BAILEY M. CARIGNAN • *Department of Biological Sciences, Plymouth State University, Plymouth, NH, USA*

ANKUR B. DALIA • *Department of Biology, Indiana University, Bloomington, IN, USA*

THOMAS O. EICHMANN • *Institute of Molecular Biosciences, University of Graz, Graz, Austria*

CLAUDIA C. HÄSE • *Department of Biomedical Sciences, College of Veterinary Medicine, Oregon State University, Corvallis, OR, USA*

JAY S. KIRKWOOD • *Department of Pharmaceutics, University of Washington, Seattle, WA, USA*

PAUL KOHL • *Institute of Molecular Biosciences, University of Graz, Graz, Austria*

JYL S. MATSON • *Department of Medical Microbiology and Immunology, University of Toledo College of Medicine and Life Sciences, Toledo, OH, USA*

KERRY MCPHAIL • *Department of Pharmaceutical Sciences, College of Pharmacy, Oregon State University, Corvallis, OR, USA*

YUSUKE MINATO • *Department of Microbiology and Immunology, University of Minnesota Medical School, Minneapolis, MN, USA*

KRISTIE C. MITCHELL • *Department of Biochemistry, Microbiology and Immunology, Wayne State University School of Medicine, Detroit, MI, USA*

ALEXANDRA E. PURDY • *Department of Biology, Amherst College, Amherst, MA, USA*

STEFAN SCHILD • *Institute of Molecular Biosciences, University of Graz, Graz, Austria; BioTechMed, Graz, Austria*

LAUREN M. SHULL • *Tufts University School of Medicine, Boston, MA, USA*

ALEKSANDRA E. SIKORA • *Department of Pharmaceutical Sciences, College of Pharmacy, Oregon State University, Corvallis, OR, USA*

ANISIA J. SILVA • *Department of Microbiology, Biochemistry and Immunology, Morehouse School of Medicine, Atlanta, GA, USA*

MIKE S. SON • *Department of Biological Sciences, Plymouth State University, Plymouth, NH, USA*

RICHARD TEHAN • *Department of Pharmaceutical Sciences, College of Pharmacy, Oregon State University, Corvallis, OR, USA*

BO R. WEBER • *Department of Pharmaceutical Sciences, College of Pharmacy, Oregon State University, Corvallis, OR, USA*
JEFFREY H. WITHEY • *Department of Biochemistry, Microbiology and Immunology, Wayne State University School of Medicine, Detroit, MI, USA*
RYSZARD A. ZIELKE • *Department of Pharmaceutical Sciences, College of Pharmacy, Oregon State University, Corvallis, OR, USA*
FRANZ G. ZINGL • *Institute of Molecular Biosciences, University of Graz, Graz, Austria*

Chapter 1

Laboratory Culturing Techniques and Maintenance of *Vibrio cholerae*

Kyle D. Brumfield, Bailey M. Carignan, and Mike S. Son

Abstract

Cholera is a severe diarrheal disease caused by the consumption of food or water contaminated with the aquatic gram-negative bacterium *Vibrio cholerae*. Infected hosts will experience vomiting and severe watery diarrhea and if not treated properly will ultimately succumb to death by dehydration. Due to the global prevalence and severity of cholera, *V. cholerae* has been extensively studied in both environmental and laboratory settings. Herein, we describe proper *V. cholerae* maintenance, in addition to classical and El Tor biotype culturing in a laboratory setting.

Key words *Vibrio cholerae*, Laboratory, Growth, Maintenance, Virulence inducing

1 Introduction

Vibrio cholerae is the aquatic gram-negative bacterium responsible for inducing the severe watery diarrhea and electrolyte loss associated with the global enteric disease cholera. This disease is endemic to many developing nations including Africa, Asia, and most recently, Central America. However, cholera also remains a significant health concern in developed nations, such as the United States, where over 100 cases were reported to the Centers for Disease Control and Prevention between 2001 and 2011; approximately 80% of these US cases were from patients who had recently traveled internationally [1]. Standard treatment for infected patients involves either oral or intravenous rehydration therapy with a regimen of antibiotics, and while vaccines for cholera do exist, they have fallen short of reaching the desired efficacy and are still under development.

Kyle D. Brumfield and Bailey M. Carignan contributed equally to this work.

Aleksandra E. Sikora (ed.), *Vibrio Cholerae: Methods and Protocols*, Methods in Molecular Biology, vol. 1839,
https://doi.org/10.1007/978-1-4939-8685-9_1, © Springer Science+Business Media, LLC, part of Springer Nature 2018

Based on the structure of the cell surface lipopolysaccharide O-antigen present, *V. cholerae* can be divided into over 200 serogroups. The O1 serogroup can be found worldwide and is responsible for all previously reported pandemics since 1817. *V. cholerae* O1 produces two main virulence factors, the toxin co-regulated pilus (TCP) and cholera toxin (CT), which allow the bacterium to colonize the host and establish infection, respectively. TCP is located on the *Vibrio* pathogenicity island (VPI) and is essential for colonization of the intestinal epithelium [2–4]. CT is located on the CTXϕ prophage and is responsible for inducing severe watery diarrhea and the subsequent electrolyte loss associated with cholera [5]. Both main virulence factors are controlled by the master virulence regulator ToxT [6, 7].

Classical and El Tor are biotypes within the O1 serogroup, which both display unique genotypic and phenotypic traits [8–10]. The classical biotype, responsible for the first six cholera pandemics, produces higher levels of CT and causes more severe disease than the El Tor biotype [11]. The current (seventh) pandemic is caused by the El Tor biotype, which has displaced the classical biotype in the environment since 1993 [12, 13] (for review *see* [14]). Recently, *V. cholerae* has evolved to demonstrate characteristics from both previously mentioned biotypes [11, 15–22], and these have since become known as El Tor variants [18, 23]. For example, a portion of the CTX operon, *ctxB*, which comprises the pentameric ring, and *tcpA*, the first gene in the TCP operon, genetically differ across the two biotypes. These distinctions can be used for reliable biotype characterization [22].

While genetic manipulation may rely on biotype distinction, culturing and maintenance are similar across both biotypes. In addition to describing how to properly grow and maintain cultures in a laboratory setting, we have also included standard protocols for inducing virulence conditions for *V. cholerae* strains of El Tor and classical biotype backgrounds, respectively. This set of protocols should leave the investigator with the knowledge to grow and maintain clinical and environmental *V. cholerae* isolates in a laboratory setting.

2 Materials

All solutions are to be prepared using ultrapure double-distilled water (prepared by purifying deionized water, to a sensitivity of 18 MΩ cm at 25 °C) and analytical-grade reagents. Unless otherwise specified, all reagents can be stored indefinitely at room temperature.

2.1 Preparation of Frozen Stocks

1. *V. cholerae* strains.
2. Luria-Bertani (LB) agar plates (standard size of 100 mm × 15 mm): weigh 5.0 g tryptone (Becton, Dickinson and Co.), 2.5 g yeast extract, 2.5 g NaCl, and 7.5 g agar, adjust volume to 500 mL with double-distilled water, and autoclave (*see* **Notes 1** and **2**).

3. LB broth: weigh 5.0 g tryptone, 2.5 g yeast extract, and 2.5 g NaCl, adjust volume to 500 mL with double-distilled water, and autoclave.
4. Sterile 60% (v/v) glycerol.
5. CryoTube™ vials.
6. −80 °C freezer.

2.2 Growth from a Frozen Stock

1. *V. cholerae* strains.
2. Luria-Bertani (LB) agar plates (standard size of 100 mm × 15 mm): weigh 5.0 g tryptone, 2.5 g yeast extract, 2.5 g NaCl, and 7.5 g agar, adjust volume to 500 mL with double-distilled water, and autoclave (*see* **Notes 1** and **2**).
3. Sterile wooden applicator sticks (Puritan).
4. 37 °C incubator.

2.3 Growth of Overnight Culture in Liquid Medium

1. *V. cholerae* strains.
2. Luria-Bertani (LB) agar plates (standard size of 100 mm × 15 mm): weigh 5.0 g tryptone, 2.5 g yeast extract, 2.5 g NaCl, and 7.5 g agar, adjust volume to 500 mL with double-distilled water, and autoclave (*see* **Notes 1** and **2**).
3. LB broth: weigh 5.0 g tryptone, 2.5 g yeast extract, and 2.5 g NaCl, adjust volume to 500 mL with double-distilled water, and autoclave.
4. Le Stab.
5. Bunsen burner.
6. 37 °C shaking incubator.

2.4 Growth Curve

1. *V. cholerae* strains.
2. Luria-Bertani (LB) agar plates (standard size of 100 mm × 15 mm): weigh 5.0 g tryptone, 2.5 g yeast extract, 2.5 g NaCl, and 7.5 g agar, adjust volume to 500 mL with double-distilled water, and autoclave (*see* **Notes 1** and **2**).
3. LB broth: weigh 5.0 g tryptone, 2.5 g yeast extract, and 2.5 g NaCl, adjust volume to 500 mL with double-distilled water, and autoclave.
4. Le Stab.
5. Bunsen burner.
6. 37 °C shaking incubator.
7. Thermo Scientific Genesys 10SUV-VIS spectrophotometer.

2.5 El Tor AKI Virulence-Inducing Conditions

1. *V. cholerae* strains.
2. Luria-Bertani (LB) agar plates (standard size of 100 mm × 15 mm): weigh 5.0 g tryptone, 2.5 g yeast extract,

2.5 g NaCl, and 7.5 g agar, adjust volume to 500 mL with double-distilled water, and autoclave (*see* **Note 2**).

3. AKI medium containing 0.03% (w/v) sodium bicarbonate. *Component 1:* weigh 7.5g Bacto™ peptone, 2.0 g yeast extract, and 2.5 g NaCl, adjust volume to 450 mL with double-distilled water, and autoclave. *Component 2:* weigh 1.5 g $NaHCO_3$, adjust volume to 50 mL of double-distilled water, and filter sterilize (*see* **Note 6**).
4. Le Stab.
5. Bunsen burner.
6. 37 °C shaking incubator.

2.6 Classical-Inducing Conditions

1. *V. cholerae* strains.
2. Luria-Bertani (LB) agar plates (standard size of 100 mm × 15 mm): weigh 5.0 g tryptone, 2.5 g yeast extract, 2.5 g NaCl, and 7.5 g agar, adjust volume to 500 mL with double-distilled water, and autoclave (*see* **Notes 1** and **2**).
3. LB broth pH 6.5: weigh 5.0 g tryptone, 2.5 g yeast extract, and 2.5 g NaCl, adjust volume to 500 mL with double-distilled water, adjust pH to 6.5, and autoclave.
4. Le Stab.
5. Bunsen burner.
6. 37 °C shaking incubator.

3 Methods

All procedures are to be carried out at room temperature unless otherwise specified. *Vibrio cholerae* is a Biosafety Level 2 (BSL-2) organism that is the causative agent of the potentially fatal intestinal disease cholera; proper handling should be enforced along with proper disposal of all waste products according to your institutional, state, and federal guidelines and/or regulations.

3.1 Preparation of Frozen Stocks

1. Prepare growth of an overnight culture in liquid media, as specified in Protocol 3 (*see* **Notes 1** and **3**). For strain selection, *see* Table 1.
2. Pellet 1.8 mL of overnight culture (≥10,000 rpm) in sterile 2 mL microcentrifuge tubes, remove supernatant by pipetting, and resuspend the cell pellet in 900 μL of fresh LB broth.
3. Transfer 900 μL resuspended culture to a sterile 2 mL CryoTube® vial.
4. Add 900 μL of sterile 60% glycerol to the culture, and mix by pipetting.
5. Label and store at −80 °C indefinitely.

Table 1
Commonly used *V. cholerae* strains

				Characteristics		
Strain	**Source**	**Background**	**NCBI accession number**	**CTXϕ**	***ctxB***	***tcpA***
O395	[22]	Classical wild-type; Ogawa	CP000626 and CP000627	Classical	Classical	Classical
N16961	[22]	El Tor wild-type; Inaba	AE003852 and AE003853	El Tor	El Tor	El Tor
C6706	[22]	El Tor wild-type; Ogawa	N/A	El Tor	El Tor	El Tor

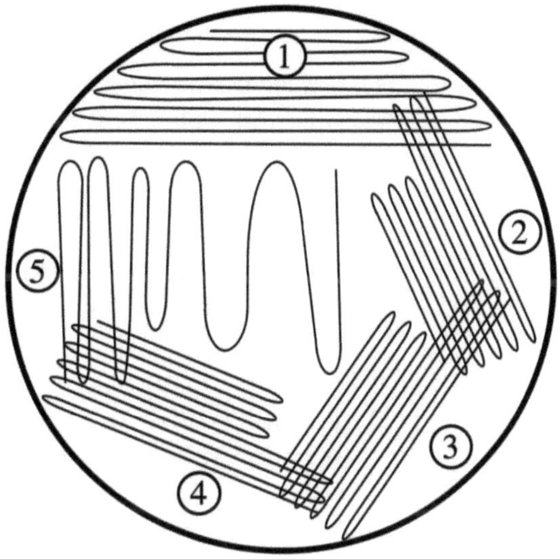

Fig. 1 Streaking pattern for single colonies. When streaking each new area, sterilize the inoculating loop using a Bunsen burner. Start in Area 1 and streak for heavy inoculation. Then, streak in Area 2, only crossing into Area 1, three to five times. Repeat this pattern for the remaining areas. In the final area, Area 5, the last streaks back and forth are also spaced out more to allow for growth of well-isolated single colonies

3.2 Growth from a Frozen Stock

1. Using a sterile applicator stick, remove a visual chunk of frozen stock culture and streak for single colonies onto LB agar plates (Fig. 1) (*see* **Note 3**).
2. Incubate plates lid-side down overnight (12–16 h) at 37 °C.

3.3 Growth of Overnight Culture in Liquid Media

1. Streak for single colonies on LB agar plates (Fig. 1), and incubate plates lid-side down overnight (12–16 h) at 37 °C (*see* **Note 3**).
2. Sterilize the inoculating wire stab using a Bunsen burner.

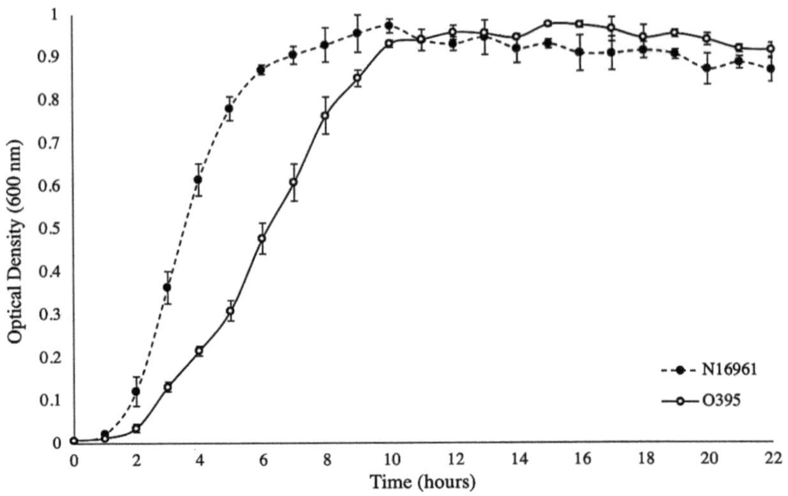

Fig. 2 *Vibrio cholerae* growth curve of El Tor wild-type N16961 and classical wild-type O395. Strains were grown in Luria-Bertani broth for up to 22 h, and the optical density was measured every hour at 600 nm

3. Inoculate LB broth using inoculating wire stab by touching a single colony and aseptically dipping the stab into liquid media.
4. Incubate with shaking overnight (12–16 h) at 37 °C.

3.4 Growth Curve

1. Prepare growth of an overnight culture in liquid media, as specified in Protocol 3 (*see* **Notes 1** and **3**).
2. Perform a 1:100 dilution by transferring 250 μL of overnight culture to 25 mL of fresh LB in a sterile 250 mL Erlenmeyer flask.
3. Incubate with shaking (24 h) at 37 °C.
4. Take the optical density at 600 nm (OD_{600}) every hour beginning at the time of inoculation (T_0) (*see* **Notes 4** and **5**).
5. Graph the OD_{600} vs. Time (Fig. 2).

3.5 El Tor AKI Virulence-Inducing Conditions

1. Streak for single colonies on LB agar plates (Fig. 1), and incubate plates lid-side down overnight (12–16 h) at 37 °C (*see* **Note 3**).
2. Sterilize the inoculating wire stab using a Bunsen burner.
3. Inoculate 10 mL AKI medium containing 0.03% $NaHCO_3$ using inoculating wire stab by touching a single colony and aseptically dipping the stab into liquid media (*see* **Note 6**).
4. Incubate without shaking (3.5 h) at 37 °C.
5. Remove and process 7 mL of culture (*see* **Note 6**).

6. Incubate the remaining 3 mL of culture with shaking (4 h) at 37 °C.
7. Process culture as needed (*see* **Note 6**).

3.6 Classical-Inducing Conditions

1. Streak for single colonies on LB agar plates (Fig. 1), and incubate plates lid-side down overnight (12–16 h) at 37 °C (*see* **Note 3**).
2. Sterilize the inoculating wire stab using a Bunsen burner.
3. Inoculate LB broth (pH 6.5) using inoculating wire stab by touching a single colony and aseptically dipping the stab into liquid media.
4. Incubate with shaking overnight (12–16 h) at 30 °C.
5. Process culture as needed.

4 Notes

1. For all assays, 1 day should be allotted for single colony growth, and an additional day is required for the growth of overnight cultures (37 °C with aeration for 12–16 h); for optimum results, overnight cultures should be used promptly (within 1 h). To perform the growth curve, a 16-h overnight starter culture is recommended.
2. Plate media can be stored indefinitely wrapped in plastic at 4 °C. All media should be stored lid-side down to prevent condensation from dripping onto the surface of the media. Prior to use of plate media, plates should be allowed to warm to room temperature.
3. Minimize the amount of time frozen stocks are out of the freezer, as repeated freeze/thaw cycles decrease cell viability.
4. It is recommended to take absorbance readings every hour until mid- to late-stationary phase and approximately every 2 hours until mid-death phase for a complete growth curve.
5. Due to the physical properties of light, results can be inaccurate with absorbance readings over 1. To avoid this, perform a 1:4 dilution of culture to media prior to reading the absorbance. A blank must be prepared prior to each absorbance reading using fresh liquid media.
6. Sodium bicarbonate ($NaHCO_3$) must be prepared immediately before use, and filter sterilized prior to being added to AKI media. Aseptically combine the two components and mix thoroughly. After 3.5 and 7.5 h, whole cell extract (WCE) can be processed for ToxT and *tcpA* analysis, respectively. After 7.5 h, cell-free supernatant can be processed for CT analysis.

Acknowledgments

Research reported in this publication was supported by an Institutional Development Award (IDeA) from the National Institute of General Medical Sciences of the National Institutes of Health under grant number P20GM103506.

References

1. Loharikar A, Newton AE, Stroika S et al (2015) Cholera in the United States, 2001–2011: a reflection of patterns of global epidemiology and travel. Epidemiol Infect 143:695–703
2. Taylor R, Miller V, Furlong D et al (1987) Use of *phoA* gene fusions to identify a pilus colonization factor coordinately regulated with cholera-toxin. Proc Natl Acad Sci U S A 84:2833–2837
3. Herrington D, Hall R, Losonsky G et al (1988) Toxin, toxin coregulated pili, and the *toxR* regulon are essential for *Vibrio cholerae* pathogenesis in humans. J Exp Med 168:1487–1492
4. Karaolis D, Johnson J, Bailey C et al (1998) A *Vibrio cholerae* pathogenicity island associated with epidemic and pandemic strains. Proc Natl Acad Sci U S A 95:31343139
5. Waldor MK, Mekalanos JJ (1996) Lysogenic conversion by a filamentous phage encoding cholera toxin. Science 272:1910–1914
6. Kovacikova G, Skorupski KS (1999) A *Vibrio cholerae* LysR homolog, AphB, cooperates with AphA at the *tcpPH* promoter to activate expression of the ToxR virulence cascade. J Bacteriol 181:4250–4256
7. Kovacikova G, Skorupski KS (2001) Overlapping binding sites for the virulence gene regulators AphB and cAMP-CRP at the *Vibrio cholerae tcpPH* promoter. Mol Microbiol 41:393–407
8. Kaper J, Bradford H, Roberts N et al (1982) Molecular epidemiology of *Vibrio cholerae* in the U.S. Gulf Coast. J Clin Microbiol 16:129–134
9. Karaolis D, Lan R, Reeves P et al (1995) The sixth and seventh cholera pandemics are due to independent clones separately derived from environmental, nontoxigenic, non-O1 *Vibrio cholerae*. J Bacteriol 177:3191–3198
10. Karaolis D, Lan R, Kaper J et al (2001) Comparison of *Vibrio cholerae* pathogenicity islands in sixth and seventh pandemic strains. Infect Immun 69:1947–1952
11. Ghosh-Banerjee J, Senoh M, Takahashi T et al (2010) Cholera toxin production by the El Tor variant of *Vibrio cholerae* O1 compared to the prototype El Tor and classical biotypes. J Clin Microbiol 48:4283–4286
12. Barua D (1992) History of cholera. In: Barua D, Greenough W III (eds) Cholera. Plenum Publishing Corporation, New York, NY
13. Morales R, Delgado G, Cravioto A (2008) Population genetics of *Vibrio cholerae*. In: Faruque S, Nair G (eds) *Vibrio cholerae—* Genomics and molecular biology. Caister Academic Press, Norfolk
14. Sack D, Sack R, Nair G et al (2004) Cholera. Lancet 363:223–233
15. Ansaruzzaman M, Bhuyian N, Nair G, et al. The Mozambique Cholera Vaccine Demonstration Project Coordination Group (2004) Cholera in Mozambique, variant of *Vibrio cholerae*. Emerg Infect Dis 10:2057–2059
16. Ansaruzzaman M, Bhuyian N, Safa A et al (2007) Genetic diversity of El Tor strains of *Vibrio cholerae* O1 with hybrid traits isolated from Bangladesh and Mozambique. Int J Med Microbiol 297:443–449
17. Lan R, Reeves P (2002) Pandemic spread of cholera: genetic diversity and relationships within the seventh pandemic clone of *Vibrio cholerae* determined by amplified fragment length polymorphism. J Clin Microbiol 40:172–181
18. Nair G, Faruque S, Bhuiyan N et al (2002) New variants of *Vibrio cholerae* O1 biotype El Tor with attributes of the classical biotype from hospitalized patients with acute diarrhea in Bangladesh. J Clin Microbiol 40:3296–3299
19. Nair GB, Safa A, Bhuiyan NA et al (2006) Isolation of *Vibrio cholerae* O1 strains similar to pre-seventh pandemic El Tor strains during an outbreak of gastrointestinal disease in an island resort in Fiji. J Med Microbiol 55:1559–1562
20. Nair GB, Mukhopadhyay AK, Safa A et al (2008) Emerging hybrid variants of *Vibrio cholerae* O1. In: Faruque SM, Nair GB (eds) *Vibrio*

cholerae—genomics and molecular biology. Horizon Scientific Press, Norwich, pp 179–190

21. Safa A, Bhuyian N, Nursin S et al (2006) Genetic characteristics of Matlab variants of *Vibrio cholerae* O1 that are hybrids between classical and El Tor biotypes. J Med Microbiol 55:1563–1569
22. Son M, Megli C, Kovacikova G et al (2011) Characterization of *Vibrio cholerae* O1 El Tor biotype variant clinical isolates from Bangladesh and Haiti, including a molecular genetic analysis of virulence genes. J Clin Microbiol 49:3739–3749
23. Nursin S, Khan G, Bhuiyan N, Ansaruzzaman M et al (2004) Diverse CTX phages among toxigenic *Vibrio cholerae* O1 and O139 strains isolated between 1994 and 2002 in an area where cholera is endemic. J Clin Microbiol 42:5854–5856

Chapter 2

Genotypic and Phenotypic Assays to Distinguish *Vibrio cholerae* Biotype

Kyle D. Brumfield, Bailey M. Carignan, and Mike S. Son

Abstract

Vibrio cholerae is a motile gram-negative bacterium found in brackish water and the etiological agent of the fecal-oral disease cholera. Classical and El Tor are two main biotypes that make up the *V. cholerae* O1 serogroup, which each display unique genotypic and phenotypic characteristics that allow for reliable biotype characterization. While treatment for cholera is much the same despite the causative strain's biotype, such classification can be imperative for laboratory experiments and may have broader impacts in the biomedical field. In the early 2000s, clinical isolates were identified that contained genotypic and phenotypic traits from both biotypes. The newly identified hybrids, termed El Tor variants, have caused clinical and environmental isolate biotype identification to be more complicated than previous single-assay identification. Herein, we describe a series of PCR-based genetic screens (*tcpA* and *ctxB*) and phenotypic assays (polymyxin B resistance, citrate metabolism, proteolytic activity, hemolytic activity, motility, and Voges-Proskauer). Together, these assays are used for reliable biotype characterization of *V. cholerae* clinical (and environmental) isolates.

Key words *Vibrio cholerae*, Biotypes, Classical, El Tor, El Tor variants, Biochemical assays

1 Introduction

Treatment of a disease and its symptoms relies on the appropriate identification of the etiological agent responsible for the disease. The causative agent of the infectious diarrheal disease cholera is *Vibrio cholerae*. Depending on the lipopolysaccharide O-antigen present, *V. cholerae* can be divided into over 200 serogroups—however, only two, O139 and O1, have endemic or pandemic potential [1, 2]. Serogroup O139 has been subject to cases of mild diarrhea, mainly in Southeast Asia [3, 4], while serogroup O1 is considered to be more clinically relevant as it has been shown to be responsible for all previously reported global pandemics since 1817.

Serogroup O1 is further divided into two biotypes: classical and El Tor. Though both biotypes demonstrate unique genotypic

Aleksandra E. Sikora (ed.), *Vibrio Cholerae: Methods and Protocols*, Methods in Molecular Biology, vol. 1839,
https://doi.org/10.1007/978-1-4939-8685-9_2,

and phenotypic traits [5–7], distinguishing between biotypes does not dictate treatment of the disease but may be necessary for biotype-dependent genetic manipulation. *V. cholerae* has evolved to demonstrate increased virulence capabilities through arising variations within current serogroups—notably the O1 serogroup. El Tor biotype strains have exhibited classical attributes and have been termed El Tor variants [8–11]. The rise of El Tor variants and expansion of virulence capabilities [10, 11] has provoked more desirable multiple-assay identification systems over the traditional single-assay strain identification. This set of protocols demonstrates genetic and phenotypic assays commonly adapted to characterize *V. cholerae* isolates [8, 12–16].

The classical biotype has been responsible for the first six cholera pandemics but has since been displaced by the El Tor biotype in the environment beginning around 1993 [17–19]. Both O1 biotypes produce two main virulence factors, the toxin co-regulated pilus (TCP) and cholera toxin (CT), which allows the bacterium to colonize within the host and is responsible for the physical manifestation of the disease symptoms, respectively. TcpA is the major pilin protein encoded by the *tcp* operon (*tcpABQCRDSTEF*) and is required for colonization of the intestinal epithelium [6, 20, 21]. CT is composed of a single CtxA subunit associated with five CtxB subunits and is responsible for inducing severe electrolyte loss and dehydration associated with cholera. Although *ctxA* is completely conserved between the two biotypes, *ctxB* and *tcpA* differ at specific base positions among the biotypes and are individually conserved within each biotype. These genetic conservations afford for reliable genetic biotype characterization. Protocol 1 demonstrates a genetic screen allowing for reliable biotype characterization of clinical (and environmental) *V. cholerae* isolates by investigating the conserved sequences of *ctxB* and *tcpA*. This protocol requires chromosomal DNA purified from overnight cultures of *V. cholerae* isolate strains and polymerase chain reaction (PCR) amplification of either *tcpA* or *ctxB* genes. Gene sequences can then be compared to reference strains (classical O395 and El Tor N16961) to help distinguish the isolate as either classical or El Tor biotype.

Phenotypic assays, including polymyxin B resistance, citrate metabolism, proteolysis, hemolysis, motility, and Voges-Proskauer (Protocols 2–7), can be used to contrast phenotypic traits of isolates in comparison to the wild-type (WT) classical strain O395, and either the WT El Tor N16961 or the WT El Tor C6706. Additionally, in this manuscript, we have included known El Tor variants MQ1795 [8], BAA-2163 [11], and BGD-5 [11] for comparison and reference purposes.

Polymyxin B is a peptide antibiotic derived from the bacterium *Bacillus polymyxa* that disrupts the outer membrane of gram-negative bacteria, making the cell more permeable and susceptible to lysis. Classical O1 biotype strains are sensitive to polymyxin B and demonstrate no growth on media supplemented with polymyxin B, while the El Tor counterparts have shown resistance to the antibiotic and can grow on said media (Fig. 1).

Citrate is an ester of citric acid and is a primary substrate of the Krebs cycle, and the ability to grow on citrate as a sole carbon source differs among biotypes. While differences in citrate metabolism among biotypes have not been extensively studied, we have consistently observed that the classical WT O395 cannot support growth on minimal citrate media. However, the El Tor biotype strains N16961 and C6706 have the ability to metabolize citrate and exhibit growth on minimal citrate media (Fig. 2).

Proteolysis assays performed on milk agar plates (casein hydrolysis) can be used to measure HapR-regulated protease activity. Although this assay does not distinguish between classical and El Tor biotypes per se, this proteolytic assay is a reliable test to help further characterize any given *V. cholerae* isolate. HapR is a promiscuous global regulator that binds to various promoters and regulates gene/operon expression [22] such as upregulation of hemagglutinin (HA)/protease [23]. Interestingly, a naturally occurring frameshift in the *hapR* gene has caused the density-dependent control of virulence gene regulation of HapR to be lost in some pathogenic strains of *V. cholerae* [22, 24]. WT classical O395 and WT El Tor N16961 strains have a deleted base in the *hapR* gene, resulting in a truncated and non-functional HapR, and are therefore phenotypically HapR negative. WT El Tor C6706, which contains a functional *hapR* gene, is said to be HapR positive visualized as an optical zone of clearance around the bacterial growth. The zone of clearance around bacterial growth can vary between different HapR positive isolates, and a larger zone of clearance is suggestive of increased HapR-regulated protease activity, or indicative of other mutations warranting further investigation (Fig. 3).

Bacteria secreting hemolytic enzymes that lyse red blood cells can be observed by spotting cultures on blood agar plates. Blood agar plates are often used to determine the degree of hemolysis between El Tor variants and can be scored as α-hemolytic (partial breakdown of hemoglobin to methemoglobin), β-hemolytic (complete lysis of red blood cells), or γ-hemolytic (no breakdown of red blood cells). WT El Tor strains N16961 and C6706 are β-hemolytic and create a zone of clearance in and around the area of bacterial growth, whereas WT classical O395, which is γ-hemolytic, does not exhibit red blood cell lysis in and around the area of bacterial growth (Fig. 4).

Motility is often associated with *V. cholerae* virulence and can be visualized using motility agar plates. Motility of isolates can be compared to the motility of WT classical O395 and El Tor N16961. When incubated under the specified conditions at 37 °C, N16961 is approximately 25% more motile than O395 (Fig. 5).

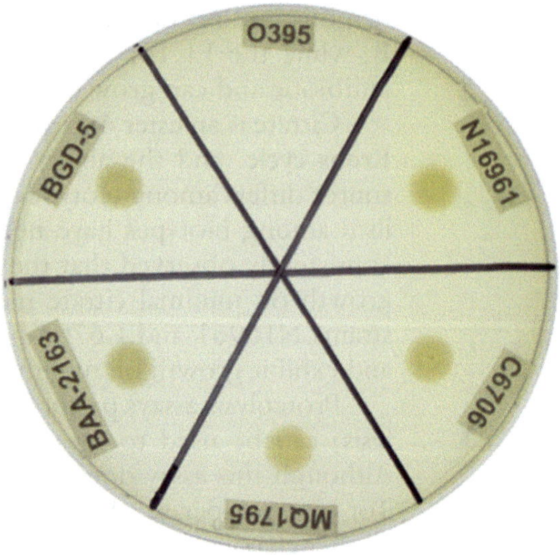

Fig. 1 Polymyxin B resistance assay. Rich medium Luria-Bertani agar plates containing 50 IU/μL polymyxin B were used to demonstrate antibiotic resistance of classical O395 (sensitive), El Tor N16961 (resistant), El Tor C6706 (resistant), as well as three El Tor variants: MQ1795 (resistant), BAA-2163 (resistant), and BGD-5 (resistant). Plates were incubated at 37 °C 7 for 18 h

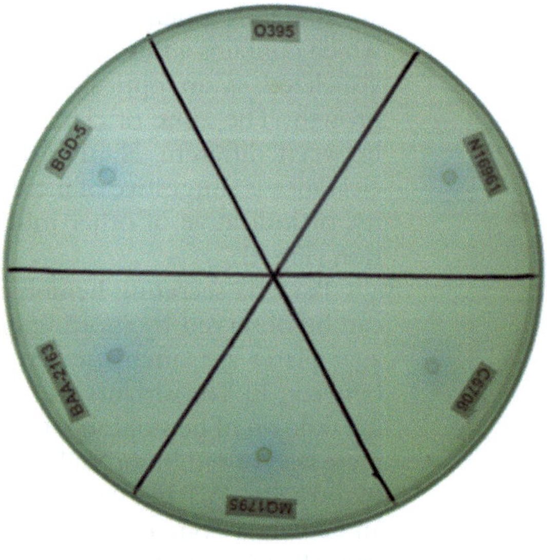

Fig. 2 Citrate metabolism assay. Minimal citrate medium plates were used to test for the ability to metabolize citrate as a sole carbon source by classical O395 (negative), El Tor N16961 (positive), El Tor C6706 (positive), and three El Tor variants: MQ1795 (positive), BAA-2163 (positive), and BGD-5 (positive). Plates were incubated at 37 °C for 24 h

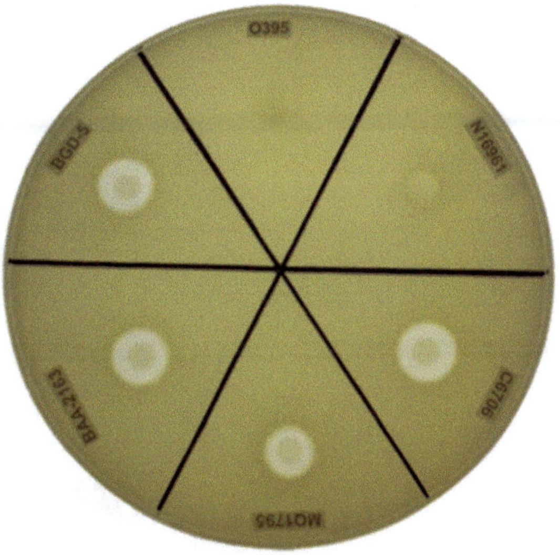

Fig. 3 Casein hydrolysis protease assay on milk agar plates. Milk agar plates were used to measure casein hydrolysis via protease activity for classical 0395 (*hapR* negative), El Tor N16961 (*hapR* negative), El Tor C6706 (*hapR* positive), and three El Tor variants: MQ1795 (*hapR* positive), BAA-2163 (*hapR* positive), and BGD-5 (*hapR* positive). Plates were incubated at 37 °C for 18 h

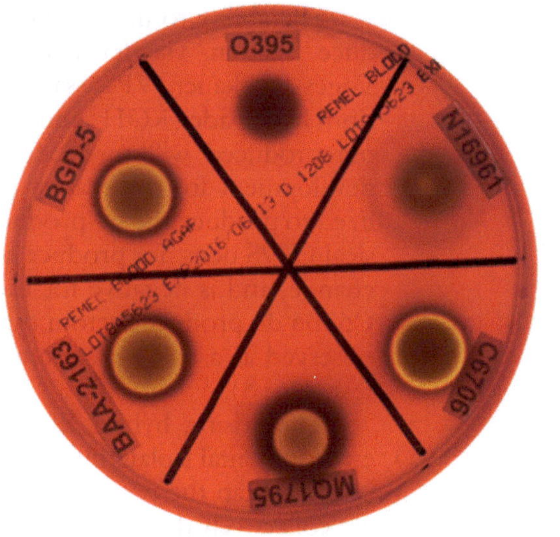

Fig. 4 Hemolysis assay on blood agar plates. Blood agar plates were used for hemolytic activity assays of classical 0395 (γ-hemolysis), El Tor N16961 (β-hemolysis), El Tor C6706 (β-hemolysis), and three El Tor variants: MQ1795 (β-hemolysis), BAA-2163 (β-hemolysis), and BGD-5 (β-hemolysis). Plates were incubated at 37 °C for 48 h

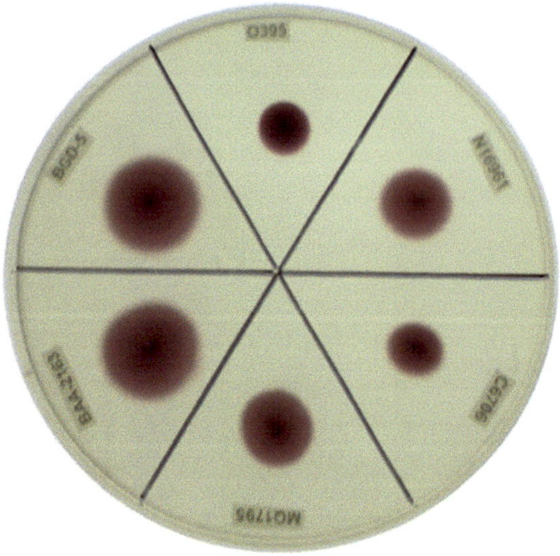

Fig. 5 Motility assay. Motility agar plates were used to measure the diameter of varying degrees of motility for the relatively less motile classical O395 (15 mm) and El Tor C6706 (15 mm) compared to the increased motility of El Tor N16961 (20 mm) and El Tor variant MQ1795 (20 mm); El Tor variants BAA-2163 (25 mm) and BGD-5 (25 mm) demonstrate hypermotility. Plates were incubated at 37 °C for 18 h

The ability to utilize glucose as a sole carbon source and produce acetoin via fermentation can be measured using the Voges-Proskauer (VP) assay. In the presence of the reagents potassium hydroxide (KOH) and alpha (α) naphthol, the resulting acetoin produced will turn deep red. Lack of the deep red color development from the assay suggests the glucose is not metabolized to produce the by-product acetoin. Using this assay, WT classical O395 does not produce acetoin, resulting in little to no color change, and is negative for the VP test. WT El Tor N16961 and C6706 do produce acetoin through glucose fermentation, and are observed as a color change to deep red, exhibiting a positive VP test (Fig. 6).

The arising hybrid El Tor variants have caused clinical and environmental isolate biotype identification to become more complicated than the previous single-assay identification. These assays are collectively designed to characterize *V. cholerae* biotypes. For all assays, we strongly encourage the use of classical WT O395 and both El Tor N16961 and C6706 (these WT El Tor strains exhibit varying phenotypic traits under the casein hydrolysis protease and motility assays) as controls. Other characterization assays are available, such as genetic screens via deep sequencing; however, they are often not necessary. The set of protocols listed here provides quick and reliable methods as an alternative to costly, labor-intensive experiments in the characterization of *V. cholerae* isolates.

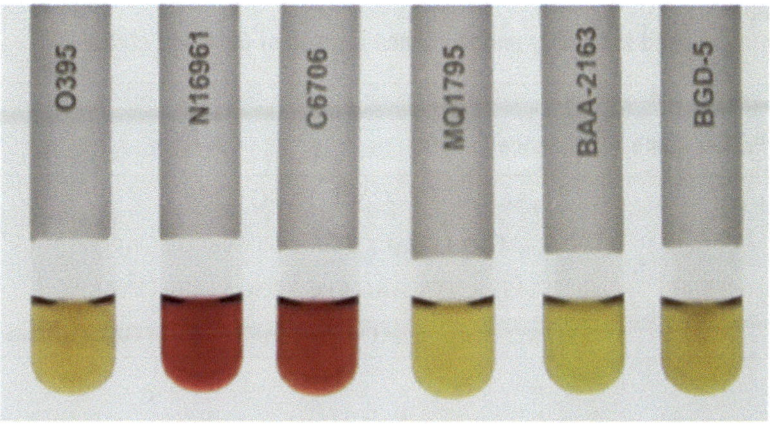

Fig. 6 Voges-Proskauer (VP) assays. Methyl red-Voges-Proskauer (MR-VP) broth was used to detect for the production of acetoin from glucose fermentation of classical O395 (negative), El Tor N16961 (positive), El Tor C6706 (positive), and three El Tor variants: MQ1795 (negative), BAA-2163 (negative), and BGD-5 (negative). Tubes were incubated at room temperature for 4 h

2 Materials

All solutions are to be prepared using ultrapure water (prepared by purifying deionized water to a sensitivity of 18 MΩ cm at 25 °C) and analytical grade reagents. Unless otherwise specified, all reagents can be stored indefinitely at room temperature.

2.1 PCR-based Genetic Screens Using *tcpA* and *ctxB*

1. *V. cholerae* strains.
2. Luria-Bertani (LB) agar plates (standard size of 100 mm × 15 mm): weigh 5.0 g tryptone, 2.5 g yeast extract, 2.5 g NaCl, and 7.5 g agar, adjust volume to 500 mL with double-distilled water, and autoclave (*see* **Note 1**).
3. LB broth: weigh 5.0 g tryptone, 2.5 g yeast extract, and 2.5 g NaCl, adjust volume to 500 mL with double-distilled water, and autoclave.
4. *tcpA*/*ctxB* forward and reverse primers (Invitrogen by Life Technologies) (Table 1).
5. Gentra Puregene Yeast/Bact. Kit (Qiagen).
6. 1.25 mM dNTP: mix 12.5 μL of 100 mM dATP (final concentration 1.25 mM), 12.5 μL of 100 mM dTTP (final concentration 1.25 mM), 12.5 μL of 100 mM dCTP (final concentration 1.25 mM), and 12.5 μL of 100 mM dGTP (final concentration 1.25 mM), and adjust volume to 1000 μL with sterile double-distilled water. Store indefinitely at −20 °C.
7. 10× standard *Taq* reaction buffer.
8. *Taq* DNA polymerase.

Table 1
Primers used to amplify and sequence regions of *tcpA* and *ctxB*

Primer name	Sequence (5′→3′)	Approximate PCR product size (bp)
tcpA-For	CCGCACCAGATCCACGTAGGTGGG	1400
tcpA-Rev	GTCGGTACATCACCTGCTGTGGGGGCAG	1400
ctxB-For	GGGAATGCTCCAAGATCATCGATGAGTAATAC	600
ctxB-Rev	CATCATCGAACCACAAAAAAGCTTACTGAGG	600

9. 1× Tris-borate-EDTA (TBE) buffer: weigh 10.8 g Tris base (final concentration 44.58 mM), 5.5 g boric acid (final concentration 44.48 mM), and 4.0 mL of 0.5 M disodium EDTA (final concentration 2.0 mM), adjust volume to 1000 mL with sterile double-distilled water.
10. 1% (w/v) agarose gel: weigh 1.0 g agarose, add 100 mL 1× TBE, and microwave at 70% power for up to 5 min (or until agarose is completely dissolved). Gels can be stored up to 48 h in 1× TBE.
11. 6× DNA gel loading dye: weigh 0.05 g bromophenol blue (final concentration 0.075 mM); add 6.0 mL of 100% glycerol [final concentration 30% (v/v)], 40 μL of 0.5 M disodium EDTA pH 8.0 (final concentration 1 mM), 1.0 mL of 1 M Tris×HCl pH 8.0 (final concentration 50 mM), and 2.0 μL of 90× gel green (final concentration 9×); and adjust volume to 20 mL with sterile double-distilled water. Store indefinitely at −20 °C.
12. DNA Clean & Concentrator™-25 kit (Zymo Research).
13. Thermo Scientific NanoDrop Lite Spectrophotometer.
14. Bio-Rad C1000 Touch™ thermal cycler.
15. 37 °C incubator with shaker.

2.2 Polymyxin B Resistance Assay

1. *V. cholerae* strains.
2. Luria-Bertani (LB) agar plates (standard size of 100 mm × 15 mm): weigh 5.0 g tryptone, 2.5 g yeast extract, 2.5 g NaCl, and 7.5 g agar, adjust volume to 500 mL with double-distilled water, and autoclave (*see* **Note 1**).
3. LB broth: weigh 5.0 g tryptone, 2.5 g yeast extract, and 2.5 g NaCl, adjust volume to 500 mL with double-distilled water, and autoclave.
4. LB agar plates containing 50 IU/mL polymyxin B (standard size of 100 mm × 15 mm): weigh 5.0 g tryptone, 2.5 g yeast

extract, 2.5 g NaCl, and 7.5 g agar, adjust volume to 500 mL with double-distilled water, autoclave, and add polymyxin B to 50 IU/mL (*see* **Notes 1** and **2**).

5. 1× phosphate-buffered saline (PBS): weigh 4.0 g NaCl, 0.1 g KCl, 0.72 g Na_2HPO_4, and 0.12 g KH_2PO_4, adjust volume to 500 mL with double-distilled water, adjust pH to 7.4, and autoclave.
6. 37 °C incubator with shaker.
7. Centrifuge ≥10,000 rpm.

2.3 Citrate Metabolism Assay

1. *V. cholerae* strains.
2. Luria-Bertani (LB) agar plates (standard size of 100 mm × 15 mm): weigh 5.0 g tryptone, 2.5 g yeast extract, 2.5 g NaCl, and 7.5 g agar, adjust volume to 500 mL with double-distilled water, and autoclave (*see* **Note 1**).
3. LB broth: weigh 5.0 g tryptone, 2.5 g yeast extract, and 2.5 g NaCl, adjust volume to 500 mL with double-distilled water, and autoclave.
4. 1× phosphate-buffered saline (PBS): weigh 4.0 g NaCl, 0.1 g KCl, 0.72 g Na_2HPO_4, and 0.12 g KH_2PO_4, adjust volume to 500 mL with double-distilled water, adjust pH to 7.4, and autoclave.
5. Minimal citrate medium agar plates (150 mm × 15 mm): weigh 7.5 g agar, 0.02 g bromothymol blue, adjust volume to 450 mL with sterile double-distilled water, autoclave, and add 50 mL of filter sterilized 10× VBMM (final concentration 1×; *see* **Note 1**).

   10× VBMM (for minimal citrate medium agar plates): weigh 1.0 g $MgSO_4 \cdot 7H_2O$, 10.0 g citric acid·H_2O, 50.0 g anhydrous K_2HPO_4, and 17.5 g $NaNH_4HPO_4 \cdot 4H_2O$, adjust volume to 500 mL with double-distilled water, and autoclave or filter sterilize using 0.2 micron filter.
6. 37 °C incubator with shaker.
7. Centrifuge ≥10,000 rpm.

2.4 Casein Hydrolysis Protease Assay on Milk Agar Plates

1. *V. cholerae* strains.
2. Luria-Bertani (LB) agar plates (standard size of 100 mm × 15 mm): weigh 5.0 g tryptone, 2.5 g yeast extract, 2.5 g NaCl, and 7.5 g agar, adjust volume to 500 mL with double-distilled water, and autoclave (*see* **Note 1**).
3. LB broth: weigh 5.0 g tryptone, 2.5 g yeast extract, and 2.5 g NaCl, adjust volume to 500 mL with double-distilled water, and autoclave.
4. 1× phosphate-buffered saline (PBS): weigh 4.0 g NaCl, 0.1 g KCl, 0.72 g Na_2HPO_4, and 0.12 g KH_2PO_4, adjust volume to

500 mL with double-distilled water, adjust pH to 7.4, and autoclave.

5. Milk agar plates (150 mm × 15 mm). *Component 1:* weigh 8.0 g instant nonfat dry milk, adjust volume to 200 mL with sterile double-distilled water, and autoclave. *Component 2:* weigh 3.68 g brain-heart infusion and 6.0 g agar, adjust volume to 200 mL with sterile double-distilled water, and autoclave (*see* **Notes 1** and **3**).
6. 37 °C incubator with shaker.
7. Centrifuge ≥10,000 rpm.

2.5 Hemolysis Assay on Blood Agar Plates

1. *V. cholerae* strains.
2. Luria-Bertani (LB) agar plates (standard size of 100 mm × 15 mm): weigh 5.0 g tryptone, 2.5 g yeast extract, 2.5 g NaCl, and 7.5 g agar, adjust volume to 500 mL with double-distilled water, and autoclave (*see* **Note 1**).
3. LB broth: weigh 5.0 g tryptone, 2.5 g yeast extract, and 2.5 g NaCl, adjust volume to 500 mL with double-distilled water, and autoclave.
4. 1× phosphate-buffered saline (PBS): weigh 4.0 g NaCl, 0.1 g KCl, 0.72 g Na_2HPO_4, and 0.12 g KH_2PO_4, adjust volume to 500 mL with double-distilled water, adjust pH to 7.4, and autoclave.
5. Blood agar (TSA with sheep blood) plates (standard size of 100 mm × 15 mm; Remel).
6. 37 °C incubator with shaker.
7. Centrifuge ≥10,000 rpm.

2.6 Motility Assay

1. *V. cholerae* strains.
2. Luria-Bertani (LB) agar plates (standard size of 100 mm × 15 mm): weigh 5.0 g tryptone, 2.5 g yeast extract, 2.5 g NaCl, and 7.5 g agar, adjust volume to 500 mL with double-distilled water, and autoclave (*see* **Note 1**).
3. LB broth: weigh 5.0 g tryptone, 2.5 g yeast extract, and 2.5 g NaCl, adjust volume to 500 mL with double-distilled water, and autoclave.
4. 1× phosphate-buffered saline (PBS): weigh 4 g NaCl, 0.1 g KCl, 0.72 g Na_2HPO_4, and 0.12 g KH_2PO_4, adjust volume to 500 mL with double-distilled water, adjust pH to 7.4, and autoclave.
5. Motility agar plates (150 mm × 15 mm): weigh 5.0 g tryptone, 2.5 g yeast extract, 2.5 g NaCl, and 1.5 g agar, adjust volume to 500 mL with double-distilled water, autoclave, and add

2.5 mL of 1% (w/v) triphenyltetrazolium chloride (TTC, *see* **Notes 1** and **4**).

Triphenyltetrazolium chloride (TTC): weigh 0.25 g triphenyltetrazolium chloride (Alfa Aesar), adjust volume to 25 mL with double-distilled water, and filter sterilize. Store wrapped in foil to protect from light.

6. Le Stab.
7. Bunsen burner.
8. 37 °C incubator with shaker.
9. Centrifuge ≥10,000 rpm.

2.7 Voges-Proskauer (VP) Assays

1. *V. cholerae* strains.
2. Luria-Bertani (LB) agar plates (standard size of 100 mm × 15 mm): weigh 5.0 g tryptone, 2.5 g yeast extract, 2.5 g NaCl, and 7.5 g agar, adjust volume to 500 mL with double-distilled water, and autoclave (*see* **Note 1**).
3. LB broth: weigh 5.0 g tryptone, 2.5 g yeast extract, and 2.5 g NaCl, adjust volume to 500 mL with double-distilled water, and autoclave.
4. Methyl red-Voges-Proskauer (MR-VP) broth: weigh 1.7 g MR-VP broth (Difco), adjust volume to 100 mL with sterile double-distilled water, and autoclave.
5. 5% (w/v) alpha (α) naphthol: weigh 1.0 g α-naphthol (MP Biomedicals), and adjust volume to 20 mL with 95% ethanol. Store up to 1 week wrapped in foil to protect from light.
6. 40% (w/v) potassium hydroxide (KOH): weigh 20.0 g potassium hydroxide (Fisher), and adjust volume to 50 mL with sterile double-distilled water. Store up to 6 months.
7. 37 °C incubator with shaker.
8. Centrifuge ≥10,000 rpm.

3 Methods

All procedures are to be carried out at room temperature unless otherwise specified. *Vibrio cholerae* is a Biosafety Level 2 (BSL-2) organism that is the causative agent of the potentially fatal intestinal disease cholera; proper handling should be enforced along with proper disposal of all waste products according to your institutional, state, and federal guidelines and/or regulations.

3.1 PCR-Based Genetic Screens Using *ctxB* and *tcpA*

1. Design primers to anneal ~50–70 bp upstream and downstream of the translational start and stop sites, respectively, of either *tcpA* or *ctxB* (*see* **Note 5**). Successful amplification and

sequencing of *tcpA* and *ctxB* have been accomplished using primers listed in Table 1.

2. Prepare overnight cultures by inoculating single colonies into rich media (*see* **Notes 6** and **7**), and isolate chromosomal DNA as per manufacturer's specifications for gram-negative bacteria (Gentra Puregene Yeast/Bact. Kit) (*see* **Note 8**).
3. Resuspend DNA pellet in resuspension buffer (EB buffer) provided in kit.
4. Ensure DNA is clean and of high quality (*see* **Note 9**) by using the Thermo Scientific NanoDrop Lite spectrophotometer.
5. For each chromosomal DNA isolate, prepare the following PCR reaction(s) on ice in individual 200 μL PCR tube(s), as itemized in Table 2:
6. Amplify templates using the following PCR parameters in a thermal cycler, as shown in Table 3:
7. Verify positive PCR products by running 5 μL of each reaction on a 1% agarose gel in 1× TBE buffer as per standard protocol (*see* **Note 10**).
8. Purify the remaining PCR products (45 μL) using DNA Clean and Concentrator™-25 kit per manufacturer's specifications (*see* **Note 11**).
9. Send PCR products for sequencing with the same primers used for PCR amplification (*see* **Note 12**).
10. Compare the isolate sequences to the published sequence available on the NCBI website (http://www.ncbi.nlm.nih.gov/gene/), and search the following accession numbers in the search query box. FASTA format of gene sequences is necessary for comparison:

Table 2
Components for standard PCR reaction

Component	Volume
Chromosomal DNA (100–200 ng)	Varies
1.25 mM dNTPs	8
10× *Taq* reaction buffer	5
Forward primer (30 pmol/μL)	1
Reverse primer (30 pmol/μL)	1
Taq polymerase	1
Sterile DDW	Up to 50 μL

Table 3
PCR parameters for thermal cycler

Step	Temp. (°C)	Time (s)
1—Initial denature	95	120
2—Denature	95	60
3—Anneal	60	45
4—Extend	72	90
5—Denature; anneal; extend Repeat steps 2 through 4 for 34 cycles	95, 60, 72	60, 45, 90
6—Final extension	72	600
7—Hold	4	∞

(a) *tcpA*: (classical) **VC0395_A0353** (El Tor) **VC_0828**

(b) *ctxB*: (classical) region between **VC0395_A1059** to **VC0395_A1060** (El Tor) **VC_1456**

3.2 Polymyxin B Resistance Assay

1. Prepare overnight cultures by inoculating a single colony into rich media (*see* **Notes 6** and **7**).
2. Wash overnight cultures three times in 1× PBS, and resuspend in 1 mL 1× PBS (*see* **Note 13**).
3. Spot 1 μL of the culture (previously washed three times in 1× PBS) on LB agar plates containing 50 IU/mL polymyxin B, and incubate the plates lid-side down overnight at 37 °C (*see* **Notes 14** and **15**).
4. After 18 h of incubation, plates can be analyzed for growth and imaged (Fig. 1).

3.3 Citrate Metabolism Assay

1. Prepare overnight cultures by inoculating a single colony into rich media (*see* **Notes 6** and **7**).
2. Wash overnight cultures three times in 1× PBS, and resuspend in 1 mL 1× PBS (*see* **Note 13**).
3. Spot 1 μL of the culture (previously washed three times in 1× PBS) on minimal citrate medium plates, and incubate the plates lid-side down overnight at 37 °C (*see* **Notes 14** and **15**).
4. After 24 h of incubation, plates can be analyzed for growth and imaged (Fig. 2).

3.4 Casein Hydrolysis Protease Assay on Milk Agar Plates

1. Prepare overnight cultures by inoculating a single colony into rich media (*see* **Notes 6** and **7**).
2. Wash overnight cultures three times in 1× PBS, and resuspend in 1 mL 1× PBS (*see* **Note 13**).

3. Spot 1 μL of the culture (previously washed three times in 1× PBS) on milk agar plates, and incubate the plates lid-side down overnight at 37 °C (*see* **Notes 14–16**).
4. After 18 h, plates can be analyzed for zones of clearance and imaged (Fig. 3).

3.5 Hemolysis Assay on Blood Agar Plates

1. Prepare overnight cultures by inoculating a single colony into rich media (*see* **Notes 6** and 7).
2. Wash overnight cultures three times in 1× PBS, and resuspend in 1 mL 1× PBS (*see* **Note 13**).
3. Spot 1 μL of the culture (previously washed three times in 1× PBS) on blood agar plates, and incubate the plates lid-side down for up to 48 h at 37 °C (*see* **Notes 14–16**).
4. After 48 h, plates can be analyzed for type and degree of hemolysis and imaged (Fig. 4).

3.6 Motility Assay

1. Prepare overnight cultures by inoculating a single colony into rich media (*see* **Notes 6** and 7).
2. Wash overnight cultures three times in 1× PBS, and resuspend in 1 mL 1× PBS (*see* **Note 13**).
3. Sterilize the inoculating wire stab using a Bunsen burner.
4. Dip the inoculating wire stab into the washed culture, and "stab" the motility agar plates vertically straight down to the bottom of the Petri dish and pull straight back up. Incubate the plates with the media-side down for 16–24 h at 37 °C (*see* **Notes 4, 14, 16,** and **17**).
5. After 16–24 h (*see* **Note 18**), plates can be analyzed by measuring the diameters of motile growth and imaged (Fig. 5).

3.7 Voges-Proskauer (VP) Assay

1. Prepare overnight cultures by inoculating a single colony into 4 mL MR-VP broth (*see* **Notes 6** and 7); alternatively, 10 μL of overnight culture may be used to inoculate 4 mL MR-VP broth (*see* **Note 13**).
2. Add 150 μL of 5% (w/v) α-naphthol and 50 μL 40% (w/v) KOH to 1 mL aliquots of each culture.
3. Vortex briefly and let stand at room temperature for up to 4 h (*see* **Note 19**).
4. After 4 h, tubes can be analyzed for color change and imaged (Fig. 6).

4 Notes

1. Plated media can be stored indefinitely wrapped in plastic at 4 °C. Milk agar plates and motility agar plates can be stored up to 1 week wrapped in plastic at 4 °C. All media should be

stored lid-side down to prevent condensation from dripping onto the surface of the media.

2. Polymyxin B can be prepared to a concentration of 50,000 IU/μL by dissolving 50 IU in 1 mL of DDW and filter sterilized using a 0.2-micron filter. Antibiotic stock solutions can be prepared separately and stored indefinitely at −20 °C. For media supplemented with polymyxin B, the antibiotic must be added once the autoclaved agar is cool to the touch, but not yet solidified. High temperatures can degrade the antibiotics.
3. Milk agar components must be prepared and autoclaved separately. Combine the two components and mix thoroughly after allowing autoclaved media to cool to the touch prior to pouring plates.
4. Plates must be poured as thick as possible (>~40 mL/plate), which is thicker than standard agar plates (~20 mL/plate).
5. When designing primers for *ctxB* and *tcpA* genes, care should be taken to reduce non-specific binding and maximize expected PCR products [25].
6. For all assays, 1 day should be allotted for single colony growth and an additional day is required for the growth of overnight cultures (37 °C with aeration for 12–16 h); for optimum results, overnight cultures should be used promptly (within 1 h).
7. It is highly recommended to include the following controls for phenotypic assays: classical O395, El Tor N16961, and El Tor C6706. Classical O395 and El Tor N16961 are not necessary to include for PCR-based genetic screens as their sequences have been published through the NCBI website. However, inclusion of these WT strains, although not necessary, can help ensure proper troubleshooting, should the PCR not work for desired samples. The sequence for C6706 is not currently published online and should be sequenced alongside the strains of interest.
8. Isolation of chromosomal DNA can be performed on overnight cultures using commercially available kits and can take up to 90 min depending on the number of samples and expertise of the individual. At this point, extracted DNA can be stored indefinitely at −20 °C.
9. Using EB buffer as a blank, clean, high-quality DNA should yield an A260/280 between 1.8 and 2.0. From a 2-mL overnight culture, resuspension of DNA in 300 μL of EB buffer can be expected to yield a concentration of 100–200 ng/μL with an A260/280 around 1.87.
10. Verification of PCR products on a 1% agarose gel should take about 90 min at 130 V, but it is recommended that PCR band migration is closely monitored.

11. Purification of the PCR product(s) can take an additional hour, and purified samples can be stored indefinitely at −20 °C.
12. Norris Cotton Cancer Center (NCCC) guidelines for PCR product sequencing preparation: in a 1.5 mL microcentrifuge tube, add (30–40 ng PCR product for tcpA and 20–30 ng PCR product for ctxB) 30–50 pmol of either forward or reverse primer in individual sequencing reactions, and bring up to 20 μL with ultrapure water.
13. To maximize efficiency, all assays (with the exception of PCR-based genetic screens) can be performed simultaneously from one single stock culture washed three times in 1× PBS. To wash cultures, pellet 1 mL of overnight culture (≥10,000 rpm) in sterile 2 mL microcentrifuge tubes, remove supernatant by pipetting, and resuspend the cell pellet in 1 mL 1× PBS; after the third wash, the culture is to remain in suspension in 1 mL 1× PBS.
14. Prior to inoculation of plated media, plates should be allowed to dry covered at room temperature for at least 24 h. Motility agar plates can take up to 48 h to dry sufficiently. Plates are considered to be sufficiently dry when there is little to no condensation on the lid of the Petri dish and no pooled liquid on the surface of the media.
15. Take care when spotting liquid cultures as splatter frequently causes cross contamination between samples. Additionally, allow for maximal spacing between spots to avoid merging the zones of clearance, and allow spots to completely soak into respective media before incubation. When spotting plates, be sure not to puncture the surface of the agar as this can negatively affect the results, giving suboptimal data.
16. Division of plates is dependent upon the number of strains being assayed; however, it is recommended to spot no more than six strains per plate. Zones of clearance can be measured in mm with a standard metric ruler.
17. When using an inoculating wire stab to inoculate motility assays, ensure wire stab does not bend during inoculation. Aggressive stabbing/inoculating may bend the stab, which may skew results.
18. Motility assays should be monitored consistently (every hour) after 16 h of incubation. This will allow for the detection of maximal motile growth prior to being overgrown.
19. After 4 h of incubation at room temperature, if strains in question do not develop a deep red or slightly pink color, they can be considered to be VP negative.

Acknowledgments

Research reported in this publication was supported by an Institutional Development Award (IDeA) from the National Institute of General Medical Sciences of the National Institutes of Health under grant number P20GM103506 and a Plymouth State University Research Advisory Council (RAC) grant for MSS.

References

1. Shimada T, Arakawa E, Itoh K et al (1994) Extended serotyping scheme for *Vibrio cholerae*. Curr Microbiol 28:175–178
2. Yamai S, Okitsu T, Shimada T et al (1997) Distribution of serogroups of *Vibrio cholerae* non-O1 non-O139 with specific reference to their ability to produce cholera toxin and addition of novel serogroups. J Jpn Infect Dis 71:1037–1045
3. Karaolis D, Lan R, Reeves P et al (1995) The sixth and seventh cholera pandemics are due to independent clones separately derived from environmental, nontoxigenic, non-O1 *Vibrio cholerae*. J Bacteriol 177:3191–3198
4. Albert J, Siddique A, Islam M et al (1993) Large outbreak of clinical cholera due to *V. cholerae* non-O1 in Bangladesh. Lancet 341:704
5. Kaper J, Bradford H, Roberts N et al (1982) Molecular epidemiology of *Vibrio cholerae* in the U.S. Gulf Coast. J Clin Microbiol 16:129–134
6. Karaolis D, Johnson J, Bailey C et al (1998) A *Vibrio cholerae* pathogenicity island associated with epidemic and pandemic strains. Proc Natl Acad Sci U S A 95:31343139
7. Karaolis D, Lan R, Kaper J et al (2001) Comparison of *Vibrio cholerae* pathogenicity islands in sixth and seventh pandemic strains. Infect Immun 69:1947–1952
8. Nair G, Faruque S, Bhuiyan N et al (2002) New variants of *Vibrio cholerae* O1 biotype El Tor with attributes of the classical biotype from hospitalized patients with acute diarrhea in Bangladesh. J Clin Microbiol 40:3296–3299
9. Nursin S, Khan G, Bhuiyan N, Ansaruzzaman M et al (2004) Diverse CTX phages among toxigenic *Vibrio cholerae* O1 and O139 strains isolated between 1994 and 2002 in an area where cholera is endemic. J Clin Microbiol 42(12):5854–5856
10. Ghosh-Banerjee J, Senoh M, Takahashi T et al (2010) Cholera toxin production by the El Tor variant of *Vibrio cholerae* O1 compared to the prototype El Tor and classical biotypes. J Clin Microbiol 48:4283–4286
11. Son M, Megli C, Kovacikova G et al (2011) Characterization of *Vibrio cholerae* O1 El Tor biotype variant clinical isolates from Bangladesh and Haiti, including a molecular genetic analysis of virulence genes. J Clin Microbiol 49:3739–3749
12. Lan R, Reeves P (2002) Pandemic spread of cholera: genetic diversity and relationships within the seventh pandemic clone of *Vibrio cholerae* determined by amplified fragment length polymorphism. J Clin Microbiol 40:172–181
13. Nair G, Safa A, Bhuyian N et al (2006) Isolation of *Vibrio cholerae* O1 strains similar to pre-seventh pandemic El Tor strains during an outbreak of gastrointestinal disease in an island resort in Fiji. J Med Microbiol 55:1559–1562
14. Ansaruzzaman M, Bhuyian N, Nair G, et al. The Mozambique Cholera Vaccine Demonstration Project Coordination Group (2004) Cholera in Mozambique, variant of *Vibrio cholerae*. Emerg Infect Dis 10:2057–2059
15. Ansaruzzaman M, Bhuyian N, Safa A et al (2007) Genetic diversity of El Tor strains of *Vibrio cholerae* O1 with hybrid traits isolated from Bangladesh and Mozambique. Int J Med Microbiol 297:443–449
16. Safa A, Bhuyian N, Nursin S et al (2006) Genetic characteristics of Matlab variants of *Vibrio cholerae* O1 that are hybrids between classical and El Tor biotypes. J Med Microbiol 55:1563–1569
17. Barua D (1992) History of cholera. In: Barua D, Greenough W III (eds) Cholera. Plenum Publishing Corporation, New York, NY
18. Morales R, Delgado G, Cravioto A (2008) Population genetics of *Vibrio cholerae*. In: Faruque S, Nair G (eds) *Vibrio cholerae—*genomics and molecular biology. Caister Academic Press, Norfolk
19. Sack D, Sack R, Nair G et al (2004) Cholera. Lancet 363:223–233

20. Taylor R, Miller V, Furlong D et al (1987) Use of *phoA* gene fusions to identify a pilus colonization factor coordinately regulated with cholera-toxin. Proc Natl Acad Sci U S A 84:2833–2837
21. Herrington D, Hall R, Losonsky G et al (1988) Toxin, toxin coregulated pili, and the *tox*R regulon are essential for *Vibrio cholerae* pathogenesis in humans. J Exp Med 168:1487–1492
22. Kovacikova G, Skorupski K (2002) Regulation of virulence gene expression in *Vibrio cholerae* by quorum sensing: HapR functions at the *aphA* promoter. Mol Microbiol 46:1135–1147
23. Jobling M, Holmes R (1997) Characterization of *hapR*, a positive regulator of the *Vibrio cholerae* HA protease gene *hap*, and its identification as a functional homologue of the *Vibrio harveyi luxR* gene. Mol Microbiol 26:1023–1034
24. Wang Y, Wang H, Cui Z et al (2011) The prevalence of functional quorum-sensing systems in recently emerged *Vibrio cholerae* toxigenic strains. Environ Microbiol Rep 3(2):218–222
25. Chuang L, Cheng Y, Yang C (2012) URPD: a specific product primer design tool. BMC Res Notes 5:306

Chapter 3

Preparation of *Vibrio cholerae* Samples for RNA-seq Analysis

Jyl S. Matson

Abstract

Massively parallel cDNA sequencing (RNA-seq) is a powerful tool for providing an unbiased approach to assess transcript abundance under a variety of conditions. In comparison to microarrays, this technique provides increased resolution and sensitivity and the ability to identify rare transcripts and sRNAs. Here, we describe the sample preparation (based on Illumina technology) used for transcriptomic analysis of *V. cholerae* cDNA libraries. We describe the entire process from RNA isolation through to the generation of barcoded cDNA libraries ready for sequencing.

Key words RNA-seq, Illumina, Next-generation sequencing, mRNA enrichment, Transcriptomic analysis, cDNA synthesis

1 Introduction

As next-generation sequencing technologies drop in price, RNA-seq has become widely used as the most common method for analyzing gene expression, as it holds many advantages over other similar techniques such as microarrays. RNA-seq allows for unbiased annotation-independent detection of transcripts, including rare transcripts and sRNAs [1, 2]. The transcript boundaries can be mapped with nucleotide resolution, resulting in the ability to identify operons, promoters, and sites of transcription attenuation [3]. Other advantages of RNA-seq include increased sensitivity and the ability to standardize output, allowing direct comparisons of results from different studies and laboratories. Importantly, RNA-seq can be successfully used under an array of conditions, including studying bacterial gene expression in animal models of infection [4, 5].

The number of biological replicates included in RNA-seq studies is crucial for obtaining meaningful results [6]. With three or more biological replicates, the power to infer differentially expressed genes in pairwise comparisons greatly increases. To allow for the sequencing of numerous biological samples at the same

Aleksandra E. Sikora (ed.), *Vibrio Cholerae: Methods and Protocols*, Methods in Molecular Biology, vol. 1839,
https://doi.org/10.1007/978-1-4939-8685-9_3, © Springer Science+Business Media, LLC, part of Springer Nature 2018

time, multiplexing should be considered. This is especially useful in bacteria, as their small genome size allows for incredibly high sequencing coverage. The addition of adaptors (barcodes) as described in this chapter allows several bacterial samples to be run in the same lane of sequencing, lowering the overall cost of analyzing additional samples and increasing the significance of the results.

In this chapter, we describe how to prepare *V. cholerae* samples for RNA-seq analysis, from RNA isolation to the point where barcoded samples are ready for commercial (or in-house) sequencing. The major steps included in this protocol are RNA isolation, removal of genomic DNA, mRNA enrichment (or rRNA depletion), cDNA synthesis, and adaptor ligation. Currently, many of these steps are available as services at sequencing facilities, removing the need for research labs to perform these steps themselves. However, in some cases there could be significant cost savings by preparing samples in the lab. Therefore, it is our intention to focus this chapter on critical aspects of sample preparation that must be carefully performed in order to generate samples of sufficient quality for RNA sequencing at any facility using Illumina technology.

2 Materials

Prepare all solutions and reactions using ultrapure water and analytical-grade reagents. Follow all appropriate waste disposal regulations when disposing waste materials. A clean workspace is required for performing experiments involving RNA. It is recommended that the work area and pipetters should be wiped down with a cleaner such as Decon ELIMINase™ before starting RNA extraction. All glassware and plasticware must be RNase-free, and gloves must be worn at all times.

2.1 RNA Extraction

1. TRIzol reagent (Invitrogen).
2. Chloroform.
3. Isopropanol.
4. 75% ethanol.
5. Qubit fluorometer (Invitrogen) or NanoDrop UV-Vis spectrophotometer (Thermo Scientific).
6. Qubit RNA BR Assay Kit.

2.2 DNase Treatment

1. Turbo DNase (Ambion).
2. Zymo RNA Clean and Concentrator 25 columns.
3. Agilent Bioanalyzer 2100.
4. Agilent RNA 6000 Nano kit.

2.3 mRNA Enrichment and Fragmentation

1. Epicenter Ribo-Zero kit for gram-negative bacteria.
2. Zymo RNA Clean and Concentrator 5 columns.
3. Affymetrix GeneChip Sample Cleanup Module.

2.4 cDNA Synthesis and Adaptor Ligation for Illumina Sequencing

1. Random primers (3 μg/μL).
2. Actinomycin D (1 μg/μL stock in RNase-free water). Store aliquots at −80 °C (*see* **Note 1**).
3. SuperScript® III reverse transcriptase first-strand synthesis kit (Invitrogen).
4. 10 mM dNTPs.
5. SUPERase-in RNase inhibitor (Ambion).
6. SuperScript® double-stranded cDNA synthesis kit (Invitrogen).
7. dUTP, PCR grade, sodium salt (Roche).
8. *E. coli* DNA ligase (10 U/μL).
9. *E. coli* DNA polymerase I.
10. 0.5 M EDTA.
11. MinElute PCR purification kit (Qiagen).
12. Illumina TruSeq RNA Sample Prep Kit (includes end repair mix, A-tailing mix, ligation mix, stop ligation buffer, resuspension buffer, and RNA adaptor indexes).
13. Agencourt AMPure XP cleanup kit.
14. Magnetic rack for 1.5 mL tubes.
15. Uracil-*N*-glycosylase (from AmpErase kit, Applied Biosystems).
16. PCR primer cocktail (Illumina).
17. Qubit dsDNA HS Assay kit.
18. Agilent DNA 1000 kit.

3 Methods

The library protocol described here was developed for sequencing on an Illumina HiSeq 2000 system. This protocol generates strand-specific information by incorporating dUTP during the second-strand cDNA synthesis [7].

3.1 RNA Extraction

Carry out all procedures on ice unless otherwise specified.

1. Grow *V. cholerae* strains under the conditions appropriate for your particular analysis (*see* **Note 2**). Centrifuge to pellet cells, 12,000 × *g* for 5 min. Resuspend cell pellet in 1 mL TRIzol and vortex to thoroughly mix sample. Incubate at room temperature for 5 min. Samples can be stored at −80 °C before proceeding to **step 2**.

2. Add 260 μL chloroform to samples. Mix vigorously, but do not vortex samples. Incubate at room temperature for 10–15 min. Centrifuge samples to separate aqueous and organic phases, 12,000 × *g* at 4 °C for 15 min.
3. Transfer upper aqueous phase (~500 μL) to a new microfuge tube and discard organic phase. Add 0.9× volume of isopropanol and vortex sample to mix (*see* **Note 3**). Incubate at room temperature for 10 min and then pellet RNA by centrifugation, 12,000 × *g* at 4 °C for 20 min.
4. Remove supernatant, being careful to not disturb the white pellet. Rinse pellet by adding 1 mL 75% ethanol, shaking the tube to mix. Centrifuge sample, 12,000 × *g* at 4 °C for 5 min.
5. Remove supernatant, again taking care to not disturb the pellet. Allow pellet to air dry until no ethanol remains. Resuspend pellet in the required amount of nuclease-free water (*see* **Note 4**).
6. Measure the RNA concentration of each sample using a Qubit or NanoDrop (*see* **Note 5**).

3.2 DNase Treatment

1. For each 50 μL reaction, use up to 5 μg RNA. Perform multiple reactions to treat a total of 20–25 μg of RNA for each sample. Each reaction should contain: 5 μL 10× Turbo DNase buffer, 2 μL Turbo DNase, up to 5 μg RNA, and enough dH_2O to total 50 μL.
2. Incubate samples at 37 °C for 30 min. After 30 min, add another 2 μL of DNase, and incubate at 37 °C for an additional 30 min.
3. Purify RNA using Zymo RNA Clean and Concentrator 25 columns according to the manufacturer's instructions.
4. Measure the resulting RNA concentration of each sample using a Qubit or NanoDrop.
5. Quality control step: verify the removal of DNA from samples by PCR. Using any primer set that amplifies a 100–500 bp sequence of chromosomal DNA, perform a standard PCR reaction using 1 μL of the cleaned RNA as the template. Check every sample at this step. If there is amplification of a PCR product, there was incomplete DNase digestion, and **steps 1–4** should be repeated for the contaminated samples before proceeding. If there are no bands, proceed to **step 6**.

   RNA samples for RNA-seq experiments must be of the highest quality. Any degradation of the ribosomal RNA will result in poor mRNA enrichment and subsequently increase the number of rRNA reads in the final data set. To ensure RNA samples are of high quality, they must be assessed on an Agilent Bioanalyzer instrument using an RNA 6000 Nano chip.

6. Following the manufacturer's instructions, assess the quality and quantity of the RNA samples by running 1 μL of each on an Agilent RNA 6000 Nano chip. Sharp peaks representing 23S and 16S rRNA should be visible. Only RNA samples with an RNA integrity number (RIN) greater than or equal to 9.8 are of a suitable quality for use in RNA-seq. Proceed to Subheading 3.3 only with those samples where the RIN >9.7.

3.3 mRNA Enrichment

1. Remove Epicenter Ribo-Zero Core kit from 4 °C, and allow tubes to thaw to room temperature. Remove Ribo-Zero rRNA removal kit from −80 °C and thaw tubes on ice.
2. Vigorously mix the RT microspheres by vortexing for 20 s to produce a homogeneous solution.
3. For each reaction, pipette 65 μL of microspheres into a separate 2 mL microsphere wash tube (do not batch wash). Avoid air bubbles and take care to pipette the full, required volume.
4. Centrifuge the dispensed microspheres, 12,000 × *g* at room temperature for 3 min. Carefully pipette off and discard the supernatant without disturbing the microsphere pellet.
5. Wash the microspheres by adding 130 μL of wash solution to each tube. Mix by vortexing at maximum speed and centrifuge, 12,000 × *g* at room temperature for 3 min. Carefully pipette off and discard the supernatant without disturbing the pellet.
6. Add 65 μL of microsphere resuspension solution, and resuspend the microspheres by vortexing at maximum speed. Add 1 μL of RiboGuard RNase inhibitor to each tube, and mix by pipetting, and/or gentle vortexing. Store the beads at room temperature until used in **step 9**.
7. Set up the Ribo-Zero rRNA removal reaction by combining the following in the order given: volume of RNA equaling 5 μg (as determined by Qubit), 4 μL Ribo-Zero reaction buffer, 10 μL Ribo-Zero rRNA removal solution, and enough RNase-free H_2O to bring the total volume of the reaction to 40 μL (*see* **Note 6**).
8. Gently mix the reaction by pipetting, and incubate at 68 °C for 10 min and then at room temperature for 15 min.
9. Mix and resuspend the prepared microspheres from **step 6**, ensuring that it is a homogenous slurry. Add the hybridized RNA sample to the resuspended microspheres in the 2 mL wash tube, and mix by pipetting 10–15 times. Mix by vortexing at medium speed, and place at room temperature.
10. Incubate the tubes at room temperature for 10 min with vortex mixing of the samples for 5 s every 3–4 min.

11. Incubate tubes at 50 °C for 10 min in a water bath or heating block. Immediately transfer the RNA-microsphere suspension to a Microsphere Removal Unit and centrifuge, 12,000 × *g* at room temperature for 1 min. Save the eluate in the collection tube and discard the filter unit.
12. Purify the mRNA from the rRNA-depleted samples using the Zymo RNA Clean and Concentrator 5 column according to the manufacturer's instructions. Elute in 12 μL RNase-free H_2O.
13. Assess the efficiency of rRNA removal by running 1 μL of each sample on an Agilent RNA 6000 Nano chip according to the manufacturer's instructions (*see* **Note 7**).

3.4 RNA Fragmentation

1. Set up the fragmentation reaction by combining the following: 8 μL 5× Affymetrix Fragmentation buffer; RNA from Subheading 3.3, **step 12**; and enough RNase-free H_2O to bring the total volume of the reaction to 40 μL.
2. Heat fragmentation reactions at 80 °C for 6 min; then cool on ice (*see* **Note 8**).
3. Purify the RNA using Zymo RNA Clean and Concentrator 5 columns according to the manufacturer's instructions. Elute in 10 μL RNase-free H_2O.

3.5 First-Strand cDNA Synthesis

1. Set up the primer/RNA template mix by combining the following in PCR tubes: 1 μL 3 μg/μL random primers; purified fragmented RNA from Subheading 3.4, **step 3**; and enough RNase-free H_2O to bring the total volume of the reaction to 10 μL.
2. Mix well by pipetting and incubate the reactions at 70 °C for 10 min in a thermocycler. Chill reactions on ice immediately for 5 min.
3. Set up the first-strand synthesis reaction by adding the following, in order, on ice: 4 μL 5× first-strand buffer, 2 μL 0.1 M DTT, 1 μL 10 mM dNTP mix, 4 μL actinomycin D, 1 μL SuperScript III (200 U/μL), and 1 μL SUPERase-in (200 U/μL) (*see* **Note 9**).
4. Mix by pipetting; then add the primer/RNA template mix from **step 2**. The reaction volume should total 23 μL. Again, mix the reaction by pipetting.
5. Incubate reaction at room temperature for 10 min; then incubate at 55 °C for 60 min in a thermocycler. Terminate the reaction by incubating the samples on ice.
6. Purify the first-strand reaction using Zymo RNA Clean and Concentrator 25 columns according to the manufacturer's instructions. Elute in 74 μL RNase-free H_2O.

3.6 Second-Strand cDNA Synthesis

1. Set up the second-strand synthesis reaction by mixing the following, in order, on ice: 4 μL 5× first-strand buffer, 2 μL 0.1 M DTT, 4 μL 100 mM dUTP, 10 μL second-strand buffer, 1 μL *E. coli* DNA ligase (10 U/μL), 4 μL *E. coli* DNA polymerase I, and 1 μL *E. coli* RNase H (*see* **Note 10**).
2. Mix by pipetting and add to purified first-strand reaction from Subheading 3.5, **step 6**. The reaction volume should total 100 μL. Again, mix the reaction by pipetting.
3. Incubate reaction at 16 °C for 2 h in a thermocycler. Place on ice immediately, and add 10 μL 0.5 M EDTA. Mix by pipetting.
4. Purify the double-stranded cDNA using the Qiagen MinElute PCR purification kit according to the manufacturer's instructions. Elute by adding 10 μL of elution buffer and letting stand for 1 min before centrifugation. Then elute with an additional 40 μL of elution buffer.

3.7 End Repair

1. Transfer the 50 μL purified double-stranded cDNA to a PCR tube.
2. Thaw an aliquot of the Illumina end repair mix, and bring to room temperature. Add 40 μL of end repair mix to each double-stranded cDNA sample, and gently pipette 10× to thoroughly mix.
3. Incubate at 30 °C for 30 min in a thermocycler.
4. Clean up the reaction using AMPure XP magnetic beads according to the manufacturer's protocol. Elute by adding 17.5 μL Illumina resuspension buffer to beads. After magnetic separation, transfer 15 μL of cleared supernatant to a PCR tube.

3.8 Adenylate 3′ Ends

1. Thaw an aliquot of Illumina A-tailing mix to room temperature.
2. Add 2.5 μL resuspension buffer to each tube of end-repaired cDNA. Pipette gently to mix thoroughly.
3. Add 12.5 μL of RT A-tailing mix to each tube. The reaction volume should total 30 μL. Mix by pipetting.
4. Incubate reaction at 37 °C for 30 min in a thermocycler. Immediately proceed to adaptor ligation.

3.9 Ligate Adaptors

Several libraries can be pooled together and sequenced on the same run. This is achieved by ligating specific adapters containing different barcode (or index) sequences to the DNA fragments. The barcodes are used to separate library-specific data after sequencing.

1. Thaw the appropriate number of Illumina adaptors for the number of samples being prepared, and bring to room tem-

perature. Also thaw an aliquot of the stop ligase mix, and bring to room temperature. Just before use, remove the DNA ligase mix from −20 °C.

2. Add 2.5 μL of DNA ligase mix to the adenylated, end-repaired cDNA. Mix by pipetting.
3. Add 2.5 μL resuspension buffer to each sample; then add 2.5 μL of a thawed RNA adapter index to each sample. Mix by pipetting (*see* **Note 11**).
4. Incubate reaction at 30 °C for 10 min in a thermocycler. Immediately add 5 μL of stop ligase mix to each sample and mix by pipetting.
5. Clean up the reaction using AMPure XP magnetic beads according to the manufacturer's protocol. Elute by adding 22.5 μL Illumina resuspension buffer to the magnetic beads. After magnetic separation, transfer 20 μL of cleared supernatant to a PCR tube.
6. Repeat **step 5**.

3.10 Uracil-N-Glycosylase Treatment

1. Add 1 μL Uracil-*N*-glycosylase to each sample. Mix by pipetting.
2. Incubate reactions at 37 °C for 15 min in a thermocycler, followed by incubation at 95 °C for 5 min. Cool reactions on ice.

3.11 PCR Enrichment

1. Thaw Illumina PCR primer cocktail and PCR master mix on ice.
2. Add 5 μL of PCR primer cocktail to each uracil-*N*-glycosylated sample. Then add 25 μL PCR master mix to each tube. Mix thoroughly by pipetting.
3. Run the following PCR program: heat to 98 °C for 30 s; then run 10 cycles of 98 °C for 10 s, 60 °C for 30 s, and 72 °C for 30 s, followed by 5 min at 72 °C and then holding at 4 °C.
4. Clean up the reaction using AMPure XP magnetic beads according to the manufacturer's protocol. Elute by adding 32.5 μL Illumina resuspension buffer to beads. After magnetic separation, transfer 30 μL of cleared supernatant to a PCR tube.
5. Quantify the DNA concentration in each sample using a Qubit or NanoDrop.
6. Following the manufacturer's protocol, assess the quality and quantity of the DNA samples by running 1 μL of each on an Agilent DNA 1000 chip. The Bioanalyzer output should show a broad peak with an average size of ~250 bp.
7. Submit the library to the sequencing facility (*see* **Note 12**).

4 Notes

1. Actinomycin is very toxic and a category 3 carcinogen. Dissolve while working in a fume hood.
2. This protocol can also be used for harvesting *V. cholerae* RNA from animal samples such as cecal fluid. However, recovery will likely be much lower than for broth cultures of bacteria.
3. The optional addition of 5 μg glycogen may be used to help pellet RNA.
4. The amount of water required for resuspension of the pellet will vary depending on the amount and type of beginning sample. In our experience, >1 μL of water will be needed for every 1 mL of starting samples from LB cultures of bacteria. It is critical to add enough water to fully resuspend the pellet before quantification.
5. Use of a Qubit fluorometer is superior to the NanoDrop for all quantification steps in this protocol. The DNA/RNA concentrations used in each kit are very sensitive, so too much or too little of added sample can cause reactions to fail. This is especially important at the adaptor ligation step.
6. The maximum amount of RNA per reaction is 5 μg; however smaller amounts can be used according to the ratios indicated in Table 1.
7. The amount of RNA remaining in each sample will be a small fraction of the starting amount; therefore, running the enriched samples on an Agilent RNA 6000 Pico kit may be necessary to obtain reliable data. The Bioanalyzer output for rRNA-depleted samples should show no discernable peaks for 5S, 16S, or 23S rRNA. If any peaks are observed corresponding in size to these rRNA species, the depletion was not successful, and samples should not be used in the subsequent RNA-seq analysis.

Table 1
Ratio of input RNA to Ribo-Zero rRNA removal solution allowed per reaction

Amount of input RNA (μg)	Maximum volume of total RNA that can be added to each reaction (μL)	Volume of Ribo-Zero rRNA removal solution per reaction (μL)
1—2.5	28	8
>2.5–5	26	10

8. At this point, it is easiest to set up reactions in PCR tubes and perform this incubation and those to follow (as indicated) in a thermocycler.
9. If multiple samples are being prepared, it is easiest to prepare a "master mix" of all reagents in this step for the total number of reactions, and then aliquot 13 μL for each individual reaction prior to adding the primer/template mixture.
10. If multiple samples are being prepared, it is easiest to prepare a "master mix" of all reagents in this step for the total number of reactions, and then aliquot 26 μL for each individual reaction prior to adding the first-strand cDNA.
11. It is critically important at this step to note which adaptor index is used for each sample. This will be used to separate library-specific data after sequencing.
12. The amount of DNA needed for sequencing is likely to vary by sequencing facility and likely to decrease over time. We suggest consultation with the sequencing facility to determine the optimal amount of library needed for submission.

References

1. Cho BK, Zengler K, Qiu Y, Park YS, Knight EM, Barrett CL, Gao Y, Palsson BO (2009) The transcription unit architecture of the *Escherichia coli* genome. Nat Biotechnol 27:1043–1049
2. Liu JM, Livny J, Lawrence MS, Kimball MD, Waldor MK, Camilli A (2009) Experimental discovery of sRNAs in *Vibrio cholerae* by direct cloning, 5S/tRNA depletion and parallel sequencing. Nucleic Acids Res 37:e46
3. Croucher NJ, Thomson NR (2010) Studying bacterial transcriptomes using RNA-seq. Curr Opin Microbiol 13:619–624
4. Mandlik A, Livny J, Robins WP, Ritchie JM, Mekalanos JJ, Waldor MK (2011) RNA-Seq-based monitoring of infection-linked changes in *Vibrio cholerae* gene expression. Cell Host Microbe 10:165–174
5. Taveirne ME, Theriot CM, Livny J, DiRita VJ (2013) The complete *Campylobacter jejuni* transcriptome during colonization of a natural host determined by RNAseq. PLoS One 8:e73586
6. Auer PL, Doerge RW (2010) Statistical design and analysis of RNA sequencing data. Genetics 185:405–416
7. Levin JZ, Yassour M, Adiconis X, Nusbaum C, Thompson DA, Friedman N, Gnirke A, Regev A (2010) Comprehensive comparative analysis of strand-specific RNA sequencing methods. Nat Methods 7:709–715

Chapter 4

Random Transposon Mutagenesis of *Vibrio cholerae*

Jyl S. Matson

Abstract

Transposon-based random mutagenesis of bacterial genomes has proven to be a powerful genetic tool for the identification of genes and regulatory elements that contribute to specific phenotypes. One such approach that has been used in *Vibrio cholerae* for many years is the introduction of mariner transposons to generate random libraries of mutants. These libraries have been successfully used for a wide variety of genetic screens and selections in this important bacterial pathogen. Here we present a detailed protocol for the use of plasmid pFD1 (containing the mariner transposon *magellan3*) to create mutant libraries in *V. cholerae*.

Key words Random mutagenesis, Transposon mutagenesis, Random-primed sequencing, Nested PCR, Mariner transposon

1 Introduction

Mariner transposons are a diverse and widespread class of transposons that are unique in that they have little host or insertion site specificity. They are relatively simple systems, requiring no other proteins or energy sources, and the only sequence requirement for insertion into recipient DNA is a TA dinucleotide. Rubin et al. were the first to show that the *Himar1* transposase, originally isolated from the horn fly *Haematobia irritans* [1], was able to transpose in a prokaryotic cell [2]. In this study, the authors constructed the plasmid that will be used in the following protocol (pFD1) and determined that the *Himar1* transposase could direct transposition in both *Escherichia coli* and *Mycobacterium smegmatis*. pFD1 is a suicide plasmid engineered to contain the minitransposon *magellan3*, harboring a gene encoding kanamycin resistance flanked by *Himar1*-inverted repeats, and the *Himar1* transposase under the transcriptional control of the *lac* promoter. Therefore, this plasmid contains all of the elements required for inducible and efficient transposition while being unable to replicate or be maintained in the target cell.

Aleksandra E. Sikora (ed.), *Vibrio Cholerae: Methods and Protocols*, Methods in Molecular Biology, vol. 1839, https://doi.org/10.1007/978-1-4939-8685-9_4, © Springer Science+Business Media, LLC, part of Springer Nature 2018

Subsequent studies in other bacteria have validated the utility of this mutagenesis strategy. pFD1 has been successfully used for screens in a wide variety of bacteria including *Caulobacter crescentus*, *Agrobacterium tumefaciens*, and *Yersinia pestis* [3–5]. In addition, modified versions of the pFD1 plasmid have been used for studies in *Pseudomonas aeruginosa*, *Campylobacter jejuni*, and *Helicobacter pylori* [6–8]. In the *Vibrio cholerae* research community, we and others have found this mutagenesis strategy to be incredibly useful for performing a variety of screens and selections, leading to a better understanding of the organism's virulence, response to stress, and ability to form biofilms [9–13].

In this chapter, we provide a simple method for random transposon mutagenesis of *V. cholerae* using plasmid pFD1. Transposon mutants can be easily selected for, due to the presence of a kanamycin cassette within the transposon sequence. In addition, we provide a detailed protocol for the identification of the transposon insertion sites in the chromosome using random-primed sequencing.

2 Materials

Prepare all solutions and reactions using ultrapure water and analytical grade reagents. Follow all appropriate waste disposal regulations when disposing waste materials.

2.1 Transposon Mutagenesis

1. *Vibrio cholerae* strain(s) to be mutagenized, streptomycin resistant (*see* **Note 1**).
2. SM10λpir *E. coli* containing plasmid pFD1.
3. LB media:10 g tryptone, 5 g yeast extract, 5 g NaCl in 1 L water. Autoclave to sterilize.
4. LB plates: LB media containing 15 g agar per L. Autoclave to sterilize, and add antibiotics before pouring plates, as appropriate.
5. Antibiotic stocks: streptomycin (100 mg/mL) and kanamycin (50 mg/mL).
6. 1 M IPTG.
7. Round nitrocellulose filter papers (e.g., Millipore, catalog #SMWP04700).

2.2 Random-Primed Sequencing

1. Primers diluted in water to a final working concentration of 30 μM.
2. PCR reagents: *Taq* polymerase (*see* **Note 2**), 10× *Taq* buffer, 25 mM $MgCl_2$, 25 mM dNTPs.
3. 1–2% agarose gel.

4. QIAquick PCR purification kit (Qiagen).
5. TE Buffer: 10 mM Tris–HCl pH 8.0, 1 mM ethylenediaminetetraacetic acid (EDTA) pH 8.0.

3 Methods

Carry out all procedures at room temperature unless otherwise specified.

3.1 Random Transposon Mutagenesis

1. Prepare overnight culture of *V. cholerae* strain to be mutagenized by inoculating one colony into LB media containing 100 μg/mL streptomycin. Prepare overnight culture of SM10λpir *E. coli* containing plasmid pFD1 by inoculating one colony into LB media containing 50 μg/mL kanamycin. Incubate overnight with shaking at 37 °C.
2. Mix together 750 μL of each overnight culture in a microfuge tube, and centrifuge to pellet cells, 12,000 × *g* for 3 min. Remove culture supernatant, and resuspend cell pellet in 50 μL of LB without antibiotics.
3. Place nitrocellulose filter on the surface of an LB agar plate (no added antibiotics) using sterile forceps. The filter will absorb moisture from the plate and become completely wet within several seconds. Place the 50 μL cell mixture on top of filter. The liquid should absorb into the filter within a few minutes.
4. Incubate plate with filter side up for 2–3 h at 37 °C.
5. Using sterile forceps, remove filter from plate, and place inside a 50 mL conical tube containing 10 mL LB, 100 μg/mL streptomycin, and 1mM IPTG to induce the transposase.
6. Vortex the conical tube in order to wash cells off the filter, and distribute them into media. Incubate the tube overnight at 37 °C with shaking.
7. Transfer the culture to microfuge tubes, and centrifuge to pellet cells, 12,000 × *g* for 3 min. Remove supernatant, and resuspend cell pellet in 10 mL LB, 100 μg/mL streptomycin, 50 μg/mL kanamycin.
8. Incubate culture for 5 h at 37 °C with shaking.
9. Plate culture on LB agar containing 100 μg/mL streptomycin and 50 μg/mL kanamycin to select for *V. cholerae* containing random transposon insertions (100 μL per plate) (*see* **Note 3**). The culture can also be plated on selective or indicator plates at this point as appropriate (*see* **Note 4**).

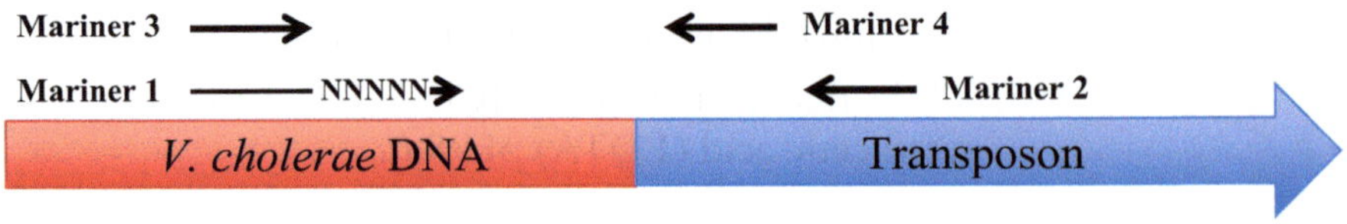

Fig. 1 Random-primed sequencing schematic. The location of the transposon insertion site in the *V. cholerae* chromosome can be identified using two sequential PCR reactions. The first reaction uses the primers designated as Mariner 1 and Mariner 2. The Mariner 1 primer contains a 3′ degenerate sequence that will allow random priming to *V. cholerae* DNA, while Mariner 2 anneals within the sequence of the transposon. The products of this reaction are used as the template in the second "nested" PCR reaction, which uses the primers designated as Mariner 3 and Mariner 4. The Mariner 3 primer has an identical sequence to the 5′ (non-degenerate) sequence of the Mariner 1 primer. The Mariner 4 primer anneals to the transposon sequence closer to the junction site than Mariner 2. This allows for a more specific amplification of the degenerate products produced by the first PCR reaction

3.2 Locating Transposon Insertion Sites by Random-Primed Sequencing

1. Prepare genomic DNA by boiling an isolated colony of *V. cholerae* containing a transposon insertion in 100 μL TE buffer for 5 min. Chill on ice for >3 min, centrifuge for 1 min at 12,000 × *g*, and then immediately remove 8 μL of the supernatant to use as the template for the first PCR reaction (Fig. 1).
2. Prepare first, random-primed PCR reaction by mixing the following, in order, on ice: 30 μL dH_2O, 5 μL 10× Taq buffer, 4 μL 25 mM $MgCl_2$, 0.5 μL Taq, 0.5 μL 25 mM dNTPs, 1 μL 30 μM primer Mariner 1 (Table 1), 1 μL 30 μM primer Mariner 2 (Table 1), and 8 μL template from **step 1**.
3. Run first, random-primed PCR program: heat to 95 °C for 8 min, then run six cycles of 95 °C for 30 s, 30 °C for 30 s, 72 °C for 1.5 min with a 5 s increase per cycle. Then run 30 cycles of 95 °C for 30 s, 45 °C for 30 s, 72 °C for 2 min with a 5 s increase per cycle, followed by 10 min at 72 °C and then holding at 4 °C.
4. Prepare second, "nested" PCR reaction by mixing the following, in order, on ice: 37 μL dH_2O, 5 μL 10× Taq buffer, 4 μL 25 mM $MgCl_2$, 0.5 μL Taq, 0.5 μL 25 mM dNTPs, 1 μL 30 μM primer Mariner 3 (Table 1), 1 μL 30 μM primer Mariner 4 (Table 1), and 1 μL of the first PCR reaction.
5. Run second, "nested" PCR program: heat to 95 °C for 8 min, then run 35 cycles of 95 °C for 45 s, 55 °C for 45 s, 72 °C for 1.5 min with a 5 s increase per cycle, followed by 10 min at 72 °C and then holding at 4 °C.
6. Separate 10 μL of each second PCR reaction by electrophoresis on a 1–2 % agarose gel. There should be one or more bands ranging in size from 50 to 2000 bp. Even if the bands are faint, proceed with sequencing. However, do not proceed to sequence any isolates that do not produce visible bands.

Table 1
Oligonucleotide sequences of the primers used for random-primed sequencing

Primer	Sequence
Mariner 1	GGCCACGCGTGCACTAGTACNNNNNNNNNNTACNG
Mariner 2	ATGCATTTAATACTAGCGACGC
Mariner 3	GGCCACGCGTGCACTAGTAC
Mariner 4	GCCATCTATGTGTCAGACCGG

7. Purify PCR products using the QIAquick PCR purification kit. At the elution step, elute the DNA using 12 μL of dH_2O, allowing it to incubate on the column 1 min before centrifugation.
8. Submit PCR products for commercial sequencing, using Mariner 4 as the sequencing primer.
9. After sequencing files are obtained, use BLAST to identify the insertion site of the transposon (*see* **Note 5**).

4 Notes

1. Most of the commonly used *V. cholerae* strains are resistant to streptomycin. However, this protocol can also be used on a streptomycin sensitive strain with minor modifications. If the strain to be mutagenized is not streptomycin resistant, purified plasmid pFD1 can be transformed into competent *V. cholerae* cells rather than introduced by mating. This eliminates the need to select against *E. coli* using streptomycin in the subsequent steps in the protocol. After transformation, proceed to induce the transposase by adding IPTG as in **step 5**. The rest of the protocol should be followed, omitting any streptomycin treatment.
2. Any brand of *Taq* polymerase can be used as long as it does not proofread.
3. May have to adjust the amount plated in order to have an appropriate number of colonies per plate, especially if using for a selection.
4. At the plating step, you have the option of using selective plates if performing a screen. You can also scrape all of the colonies from multiple plates to pool in order to generate a mutant library. Libraries can be stored as −80 °C freezer stocks (add glycerol to a final concentration of 20%) for future screen-

ing, or individual colonies can be isolated and stored in 96-well plates (again, add glycerol if stocks are to be maintained long term).

5. In our experience, not every reaction will produce sequencing results. However, we typically get satisfactory results from approximately 80–90% of sequencing reactions. The exact location of the junction between transposon and chromosomal sequence is not always apparent from the initial sequencing run as the Mariner 4 primer anneals close to the end of the transposon. Additional sequencing using primers specific to the gene/region of interest can be used to validate the exact junction with the transposon if it is not achieved with the initial sequencing reaction.

References

1. Lampe DJ, Churchill ME, Robertson HM (1996) A purified mariner transposase is sufficient to mediate transposition in vitro. EMBO J 15:5470–5479
2. Rubin EJ, Akerley BJ, Novik VN, Lampe DJ, Husson RN, Mekalanos JJ (1999) In vivo transposition of mariner-based elements in enteric bacteria and mycobacteria. Proc Natl Acad Sci U S A 96:1645–1650
3. Curtis PD, Klein D, Brun YV (2013) Effect of a *ctrA* promoter mutation, causing a reduction in CtrA abundance, on the cell cycle and development of *Caulobacter crescentus*. BMC Microbiol 13:166
4. Heckel BC, Tomlinson AD, Morton ER, Choi JH, Fuqua C (2014) *Agrobacterium tumefaciens exoR* controls acid response genes and impacts exopolysaccharide synthesis, horizontal gene transfer, and virulence gene expression. J Bacteriol 196:3221–3233
5. Pan NJ, Brady MJ, Leong JM, Goguen JD (2009) Targeting type III secretion in *Yersinia pestis*. Antimicrob Agents Chemother 53: 385–392
6. Guo BP, Mekalanos JJ (2001) *Helicobacter pylori* mutagenesis by mariner in vitro transposition. FEMS Immunol Med Microbiol 30:87–93
7. Hendrixson DR, Akerley BJ, DiRita VJ (2001) Transposon mutagenesis of *Campylobacter jejuni* identifies a bipartite energy taxis system required for motility. Mol Microbiol 40:214–224
8. Wong SM, Mekalanos JJ (2000) Genetic footprinting with mariner-based transposition in *Pseudomonas aeruginosa*. Proc Natl Acad Sci U S A 97:10191–10196
9. Matson JS, DiRita VJ (2005) Degradation of the membrane-localized virulence activator TcpP by the YaeL protease in *Vibrio cholerae*. Proc Natl Acad Sci U S A 102: 16403–16408
10. Matson JS, Yoo HJ, Hakansson K, DiRita VJ (2010) Polymyxin B resistance in El Tor *Vibrio cholerae* requires lipid acylation catalyzed by MsbB. J Bacteriol 192:2044–2052
11. Hase CC, Mekalanos JJ (1999) Effects of changes in membrane sodium flux on virulence gene expression in *Vibrio cholerae*. Proc Natl Acad Sci U S A 96:3183–3187
12. Van Dellen KL, Houot L, Watnick PI (2008) Genetic analysis of *Vibrio cholerae* monolayer formation reveals a key role for DeltaPsi in the transition to permanent attachment. J Bacteriol 190:8185–8196
13. Minato Y, Fassio SR, Wolfe AJ, Hase CC (2013) Central metabolism controls transcription of a virulence gene regulator in *Vibrio cholerae*. Microbiology 159:792–802

Chapter 5

Metabolomics of *Vibrio cholerae*

Yusuke Minato, Jay S. Kirkwood, and Claudia C. Häse

Abstract

Metabolomics is an "omics" approach to quantitatively measure a large set of metabolites. In this chapter, we describe an example method for performing liquid chromatography coupled to mass spectrometry (LC–MS)-based untargeted metabolomics on a cell extract from *Vibrio cholerae*.

Key words Untargeted metabolomics, Liquid chromatography–mass spectrometry (LC–MS), Phenyl-bonded phase, Reversed-phase chromatography, Q-TOF mass spectrometer

1 Introduction

Clearly, the metabolism of pathogenic microorganisms has to be flexible as the efficient replication of pathogenic bacteria in the host is the outcome of an active adaptation process. *Vibrio cholerae* is exposed to a variety of environments within the human host and eventually has to adapt to the nutrients available at its preferred infection site, the small intestine. The small intestine is the part of the gastrointestinal tract where the majority of digestion and absorption of food takes place. As the bacteria start to cause the profuse diarrhea that is the hallmark of the disease, they have to once again adapt to the resulting changes in nutrient availability to continue the infection process. It can therefore be expected that the metabolic status in *V. cholerae* changes substantially during the course of the infection and that changes in metabolism are coordinated with virulence gene expression.

The major aim of metabolomics is the quantitative analysis of the collection of small molecules that exist in a given system. Several platforms are routinely used for metabolomics analysis including nuclear magnetic resonance (NMR), gas chromatography coupled to mass spectrometry (GC–MS), and liquid chromatography coupled to MS (LC–MS). LC–MS is perhaps the most widely used technique for metabolomics studies and is the focus of this protocol.

Aleksandra E. Sikora (ed.), *Vibrio Cholerae: Methods and Protocols*, Methods in Molecular Biology, vol. 1839, https://doi.org/10.1007/978-1-4939-8685-9_5,

Metabolomics studies can usually be classified as targeted or untargeted. With a targeted approach, a limited panel of previously characterized small molecules is optimized for detection before sample analysis. This is typically done with a triple quadrupole mass spectrometer operating in selected reaction monitoring mode. While the targeted approach is a sensitive one, limitations are imposed on discovery as only known metabolites can be analyzed. Untargeted metabolomics aims to quantitatively analyze all metabolites in a sample within a given mass range, usually between 50 and 1200 Da. Untargeted experiments are typically performed using high resolving power instruments, such as time-of-flight (TOF) and Orbitrap mass spectrometers. Though not as sensitive as targeted approaches, untargeted metabolomics studies are less biased and have the potential to uncover novel metabolites.

Over the past decade, metabolomics has evolved into a powerful strategy to probe cellular metabolism and has helped reshape our understanding of microbial metabolism. In this chapter, we describe an example method for performing LC–MS-based untargeted metabolomics on a cell extract from *V. cholerae*. In this protocol, we specifically grow *V. cholerae* in a cholera toxin inducing growth condition, but the protocol is adaptable for other cholera toxin inducing and non-inducing growth conditions.

2 Materials

Prepare all solutions using LC–MS-grade water and analytical grade reagents.

2.1 Metabolite Extraction

1. Growth media (*see* **Note 1**).
2. Table top refrigerated centrifuge.
3. Centrifuge tube (1.5 mL).
4. Extraction buffer: 1 mM HEPES and 1 mM EDTA (pH 7.2) in ice-cold LC–MS-grade water.
5. Liquid nitrogen.
6. Heating block.

2.2 LC–MS

1. HPLC system with cooled autosampler (*see* **Note 2**).
2. Stationary phase heating chamber.
3. Phenyl-3 stationary phase (*see* **Note 3**) (Inertsil phenyl-3, 150 × 4.6 mm, 5 μM).
4. Buffer A: Add 1 mL of LC–MS grade formic acid to 999 mL of LC–MS-grade water (*see* **Note 4**).

5. Buffer B: Add 1 mL of LC–MS-grade formic acid to 999 mL of LC–MS-grade MeOH.
6. Q-TOF mass spectrometer (*see* **Note 5**).
7. Calibrant delivery system (AB SCIEX).
8. Calibration solutions (AB SCIEX).
9. Data processing software (*see* **Note 6**).
10. PeakView data analysis software.

3 Methods

LC–MS-based untargeted metabolomics workflow includes sample preparation, LC–MS/MS, data analysis, and metabolite identification (Fig. 1).

3.1 Sample Preparation

1. Grow bacterial culture under conditions of interest (*see* **Note 1**).
2. Harvest the cells by centrifugation (8000 rpm, 4 °C, 5 min) (*see* **Notes 7** and **8**).
3. Carefully remove the supernatant by aspiration.

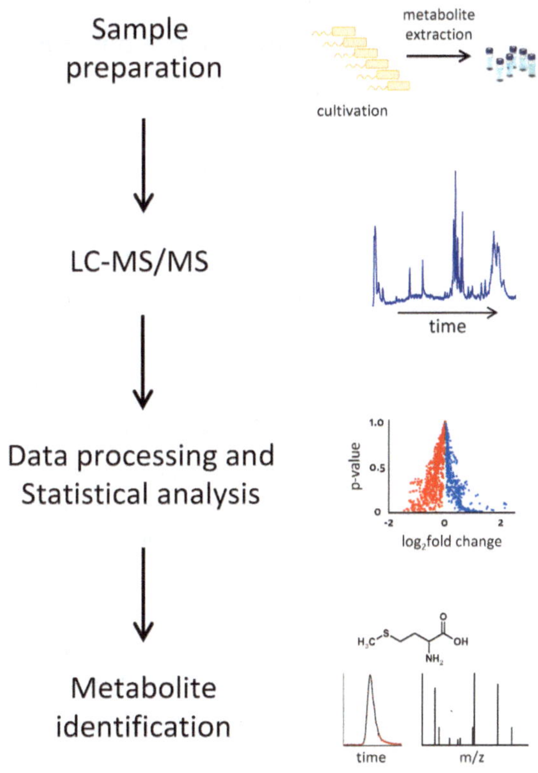

Fig. 1 A workflow for untargeted metabolomics that includes sample preparation, LC–MS/MS, data analysis, and metabolite identification

4. Resuspend the bacterial culture in 600 μL of ice-cold LC–MS-grade water in 1 mM HEPES and 1 mM EDTA (pH 7.2) by vortexing for 30 s (*see* **Note 9**).
5. Incubate the sample in boiling water for 2 min.
6. Incubate the sample in liquid nitrogen for 1 min.
7. Thaw the sample at room temperature.
8. Repeat **steps 6** and 7.
9. Incubate the sample at −20 °C for 1 h.
10. Remove the cell debris by centrifugation (16,200 × *g*, 4 °C, 15 min).
11. Filtrate the supernatant (0.22 μm pore size).
12. Transfer the filtrate to a HPLC vial and store at −80 °C.

3.2 LC–MS

1. Equilibrate the HPLC system and stationary phase to the following conditions:

   95% buffer A and 5% buffer B

   400 μL/min

   50 °C column oven

   4 °C autosampler temperature

2. Set the MS conditions as follows (*see* **Note 10**):

   Source temperature 500 °C

   IonSpray Voltage Floating 4500 V

   Curtain gas 25

   Declustering potential 80

   Collision energy 40

   Collision energy spread 15

   MS acquisition time 0.25 s

   MS scan range 40–1200 m/z

   MS/MS acquisition time 0.17 s

   MS/MS scan range 40–1200 m/z

3. Prepare a quality control (QC) sample: Pool together an equal volume of all samples into a single vial to be analyzed throughout the LC–MS analysis.
4. Set the injection volume to 10 μL, and configure the calibrant delivery system to automatically perform mass calibrations before the injection of every sample. Calibration adds 1–2 min to the total run time.
5. Start the analysis of a sample batch by injecting two blanks (blank MeOH) followed by triplicate analysis of the QC

sample. Confirm reproducibility of total ion signal and pressure trace for the triplicate injections of the QC sample.

Chromatographic gradient is as follows:

Time in minutes, t

$t = 0$, 5% B

$t = 5$, 5% B

$t = 11$, 30% B

$t = 23$, 100% B

$t = 35$, 100% B

$t = 37$, 5% B

$t = 50$, 5% B

6. Continue with the analysis of samples in a randomized order, injecting the QC sample every 3–6 samples to monitor system stability over the course of the analysis.
7. Perform **steps 1–6** in positive and negative ionization modes to increase metabolome coverage.
8. Data processing includes peak detection, peak alignment, peak integration, and data normalization across all injections, including the QC sample. Here we use MarkerView software (AB SCIEX), but other options are available (*see* **Note 6**).

3.3 Data Analysis and Metabolite Identification

1. Normalize each sample to total ion signal to account for analytical variability associated with sample extraction and LC–MS analysis.
2. Determine which features are of interest (*see* **Note 11**).
3. Search for the m/z value of a given feature in online databases (METLIN, HMDB, MassBank). Generally, the greater abundance features generate more accurate mass measurements. The mass tolerance when matching features to a database will thus depend on abundance, as well as MS instrumentation. In this protocol, we ignore matches > 10 ppm error.
4. Compare the experimental MS spectrum for a feature to potential matches found in **step 3** (*see* **Note 12**). If MS/MS was not acquired on a feature of interest, a subset of samples can be reanalyzed in product ion mode to acquire this data.
5. Once a feature has been identified as a metabolite by mass and MS/MS fragmentation in **steps 3** and **4**, the retention time should be compared to an authentic standard. The retention time tolerance depends on the LC system conditions as well as the metabolite. With this protocol, we tolerate a 15 s difference in retention time.
6. Report the parameters used for metabolite identification and the associated error.

4 Notes

1. We grew *V. cholerae* O395N1 strain in 5 mL of LB (pH 6.5) at 30 °C for 2 h aerobically, known as a cholera toxin induction condition for *V. cholerae* O1 classical biotype [1]. However, it is better to use a chemically defined media if possible.
2. For this protocol, we used a Shimadzu Nexera system; however, any HPLC or UPLC system that is stable at low back pressure, about 400 psi, is suitable.
3. For this protocol, a phenyl-3 stationary phase was used as it retains fairly well both polar and nonpolar metabolites [2]. Different stationary phases can be used to better target either polar metabolites (i.e., zwitterionic-HILIC [3]) or nonpolar metabolites (i.e., C18 [4]).
4. Metabolite retention and separation depends largely on the LC buffers of choice, in addition to the stationary phase. Different buffer systems can be used to increase the separation of different metabolite classes. Here, we have optimized the buffer system specifically for the phenyl-3 stationary phase, placing priority on the retention of polar metabolites.
5. This protocol was carried out using a Triple TOF 5600 equipped with a TurboSpray electrospray ionization source; however, any Q-TOF mass spectrometer, ideally with data-dependent MS/MS acquisition capability, is suitable.
6. Here we use MarkerView software (AB SCIEX); however, vendor-independent, open source tools for processing LC–MS metabolomics data, such as XCMS [5], are frequently used.
7. When necessary, more rapid quenching can be achieved by collecting cells by membrane filter (0.45 μm) and freezing in liquid nitrogen.
8. To obtain similar amount of cell extract from each bacterial culture, we determined the amount of bacterial liquid culture to be harvested based on the density of bacterial culture (determined by measuring $O.D_{600}$).
9. Different solvents can be used to increase the extraction of different metabolite classes [6].
10. The settings described here are specific to the 5600 mass spectrometer. Mass spectrometers from other vendors may require different settings.
11. In the context of untargeted metabolomics, a feature is defined as any m/z value at a unique retention time or a unique m/z value at any retention time. Due to in-source fragmentation, dimerization, and adduction, a single metabolite is often represented by a collection of features. The deter-

mination of which features are of interest to a given experiment is most often driven by statistics. In the case where two experimental groups exist, potentially important features can be extracted by a two-sided *t*-test with correction for multiple testing. Features with the lowest p-values from this test would represent features of interest. If a more targeted question is being asked, a targeted data analysis can be performed on the untargeted LC–MS data. For example, the features that represent all intermediates from the TCA cycle can be manually extracted and quantified in each sample.

12. There is currently no well-accepted set of criteria for the matching of experimental MS/MS spectra to those in online databases. Confirmation of metabolite identity with an authentic standard should be performed to increase confidence.

5 Conclusions

Several studies support the notion that metabolism and virulence are closely related in several pathogenic bacteria including *V. cholerae* [7]. Metabolomics is a relatively new but powerful "omics" technology that can capture a global snapshot of *V. cholerae* metabolism that might allow us to understand how metabolism and virulence are linked in this organism.

Acknowledgments

We thank Dr. Fred Stevens for helpful discussions. This research was supported by a grant from the National Institutes of Health to C.C.H. [AI-063121-02].

References

1. Minato Y et al (2014) Roles of the sodium-translocating NADH:quinone oxidoreductase (Na+-NQR) on vibrio cholerae metabolism, motility and osmotic stress resistance. PLoS One 9:e97083
2. Kirkwood JS, Maier C, Stevens JF (2013) Simultaneous, untargeted metabolic profiling of polar and nonpolar metabolites by LC-Q-TOF mass spectrometry. Curr Protoc Toxicol Chapter 4, Unit 4:39
3. Zhang R et al (2014) Evaluation of mobile phase characteristics on three zwitterionic columns in hydrophilic interaction liquid chromatography mode for liquid chromatography-high resolution mass spectrometry based untargeted metabolite profiling of Leishmania parasites. J Chromatogr A 1362:168–179
4. Sheikh KD, Khanna S, Byers SW, Fornace A, Cheema AK (2011) Small molecule metabolite extraction strategy for improving LC/MS detection of cancer cell metabolome. J Biomol Tech 22:1–4
5. Smith CA, Want EJ, O'Maille G, Abagyan R, Siuzdak G (2006) XCMS: processing mass spectrometry data for metabolite profiling using nonlinear peak alignment, matching, and identification. Anal Chem 78:779–787
6. Want EJ et al (2006) Solvent-dependent metabolite distribution, clustering, and protein extraction for serum profiling with mass spectrometry. Anal Chem 78:743–752
7. Rohmer L, Hocquet D, Miller SI (2011) Are pathogenic bacteria just looking for food? Metabolism and microbial pathogenesis. Trends Microbiol 19:341–348

Chapter 6

Natural Cotransformation and Multiplex Genome Editing by Natural Transformation (MuGENT) of *Vibrio cholerae*

Ankur B. Dalia

Abstract

Generating mutant strains is an essential component of microbial genetics. Natural genetic transformation, a process for the uptake and integration of foreign DNA, is shared by diverse microbial species and can be exploited for making mutant strains. Canonically, this process has been used to generate single mutants and sequentially to generate strains with multiple mutations. Recently, we have described a method for multiplex genome editing by natural transformation (MuGENT), which allows the generation of strains with multiple scarless mutations in a single step. Here, we provide a detailed description of the methods used for mutagenesis of the cholera pathogen *Vibrio cholerae* with a particular emphasis on mutagenesis via MuGENT.

Key words *Vibrio cholerae*, Natural transformation, Multiplex genome editing, Combinatorial mutagenesis, Accelerated evolution, Cotransformation

1 Introduction

The ability to generate mutant strains is essential for microbial genetics. Diverse microbial species exhibit natural competence, a physiological state in which these microbes can take up DNA from their extracellular environment [1]. If homologous to the host chromosome, this ingested DNA can be integrated into the genome by RecA-mediated homologous recombination. The cumulative process of DNA uptake and integration is known as natural transformation. Since this natural process can be used to alter the host genome, natural transformation can be exploited in a lab setting to generate mutant strains for genetic studies.

In 2006, Meibom et al. discovered that in *V. cholerae*, the causative agent of the diarrheal disease cholera, chitin oligosaccharides induce competence and transformation [2]. Other groups have gone on to demonstrate that chitin also induces competence in other members of the *Vibrionaceae* [3–5]. When a naturally competent population of *V. cholerae* is given transforming DNA (tDNA), only a small fraction of cells (i.e., ~10^{-2}–10^{-5}) will

Aleksandra E. Sikora (ed.), *Vibrio Cholerae: Methods and Protocols*, Methods in Molecular Biology, vol. 1839,
https://doi.org/10.1007/978-1-4939-8685-9_6,

successfully integrate this tDNA into their chromosome [2, 6–8]. Thus, traditionally, natural transformation is used to make mutant strains via the use of tDNA with a selectable marker (i.e., an antibiotic resistance cassette flanked by regions of homology that dictate the locus to be mutated). By pairing natural transformation with site-specific recombinases and engineered antibiotic resistance cassettes (i.e., FRT/Flp), a two-step procedure can be used to make in-frame mutants that have a "scar" sequence which remains following site-specific recombination of the antibiotic resistance marker [6, 9, 10]. This process is not, however, capable of making scarless mutations, which is essential for generating defined deletions, insertions, and point mutations. Alternative procedures for scarless mutagenesis utilize selectable and counter-selectable markers; however, these methods still require two steps [11].

Previous studies have demonstrated that markers can be cotransformed into a population of naturally competent cells at frequencies higher than predicted based on the integration of each marker independently [12]. This suggests that cells that are capable of becoming transformed within a population are more likely to take up and integrate multiple fragments of tDNA.

Recently, we have optimized the protocol for natural cotransformation of genetically unlinked PCR products in *V. cholerae* [13]. For cotransformation, one PCR product has an antibiotic resistance marker (selected product), while the other PCR product has the unmarked mutation of interest (unselected product). Transformations are then plated to select for integration of the selected product and screened to assess the integration of the unselected product. Using this principle, we have demonstrated that point mutations, deletions, and insertions of varying size can be integrated (as the unselected product), in a single step, into the genome at efficiencies of up to 50%. Furthermore, we showed that natural cotransformation allows for the simultaneous integration of multiple mutations (genome edits) into the genome in a single step. So, we have called this method MuGENT for *mu*ltiplex *g*enome *e*diting by *n*atural *t*ransformation [13]. This process can also be performed in multiple iterative cycles to increase the number and complexity of genome edits that can be integrated into a bacterial population (Fig. 1). Below, we describe in detail the methods used for MuGENT in *V. cholerae*.

2 Materials

2.1 Design and Construction of Selected and Unselected Products

1. High-fidelity polymerase PCR reagents (Phusion, ThermoFisher).
2. PCR purification kit (QIAquick PCR Purification Kit, Qiagen).
3. Microvolume spectrophotometer for A_{260}/A_{280} measurements (NanoDrop, ThermoFisher).

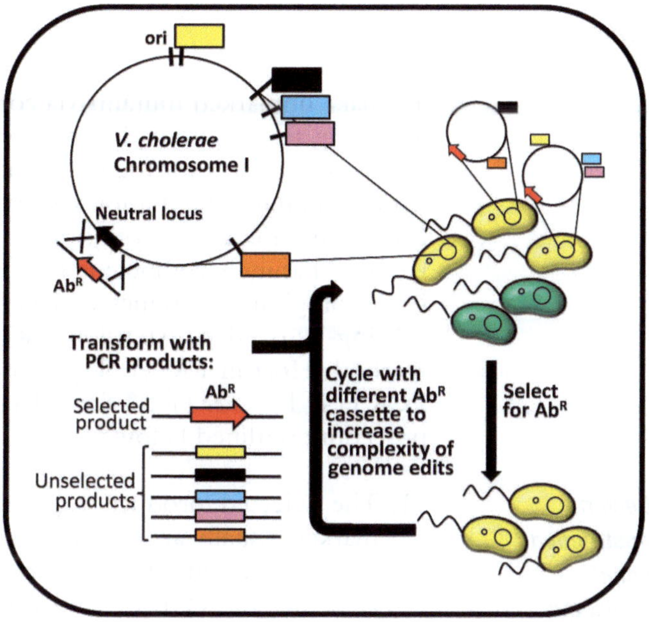

Fig. 1 Schematic demonstrating how multiple mutations can be simultaneously and iteratively introduced into a population of cells using MuGENT. Colored boxes signify distinct unselected genome edits

2.2 MuGENT of Naturally Competent V. cholerae

1. Spectrophotometer for OD_{600} measurement.
2. Instant Ocean (IO) = 7 g/L instant ocean sea salts (Instant Ocean) in milliQ H_2O. Autoclave or filter through a 0.2 μM filter to sterilize. This solution can also be made as a 4× stock (i.e., 28 g/L instant ocean sea salts in milliQ H_2O) and then diluted to 1× with sterile H_2O.
3. Chitin slurry = Take 8 g of chitin powder (Alfa Aesar), and add to 150 mL of IO. Autoclave to sterilize.
4. Kanamycin, spectinomycin, trimethoprim, and/or other appropriate antibiotics.

2.3 Assessing Complexity of Genome Edits in the Mutant Population and/or Isolating Colonies with Desired Genome Edits

1. *Taq* polymerase (2000 U/mL) (Syd Labs).
2. 5× *Taq* reaction buffer = 100 mM KCl, 200 mM Tris–HCl pH 8.8, 100 mM $(NH_4)_2SO_4$, 30 mM $MgSO_4$, and 1% Triton X-100.
3. 80% glycerol (Sigma Aldrich).

3 Methods

To make unmarked mutants via cotransformation/MuGENT, the following are required: [1] a selected product (i.e., a PCR product that you can select for the integration of), [2] an unselected product containing the mutation you would like to generate flanked by 3 Kb of homology on both sides of the mutation, and [3] a population of naturally competent *V. cholerae*. A familiarity with general bacteriological techniques and generating mutant constructs by splicing-by-overlap extension (SOE) PCR is assumed for this method. Horton RM et al. provide a detailed protocol for SOE PCR [14, 15]. *See* Fig. 2 for a detailed diagram of the MuGENT procedure outlined below.

3.1 Design and Construction of Selected and Unselected Product

1. The selected product is designed to integrate a selectable marker (usually an antibiotic resistance cassette) into a neutral locus in the genome (*see* **Note 1**). Neutral loci used in *V. cholerae* include VC1807 and *lacZ* because disruption of these genes has not been shown to confer any fitness disadvantage to the organism. Other neutral loci that can be used include intergenic regions of the genome between convergently transcribed genes. There are no additional constraints on the genetic locus for the selected product (i.e., distance from the unselected product does not affect cotransformation efficiency). The length of homology on each side of the mutation in the selected product is not critical and can range from 1 to 3 Kb. PCR products with shorter arms of homology will reduce the total number of transformants obtained during the cotransformation procedure.
2. The unselected product requires 3 Kb of homology on each side of the mutation for maximum cotransformation efficiencies. Any type of genome edit can be designed into the unselected product (i.e., site-specific point mutations, domain deletions, insertions, promoter swaps, randomized nucleotides for ribosome binding site optimization, error-prone PCR mutagenized products, etc.). Point mutations and deletions/insertions of ≤500 bp can be integrated at cotransformation efficiencies of up to 50%, while deletions and insertions of >500 bp may be integrated at lower rates.
3. All PCR amplifications for construction of selected and unselected products are performed using a high-fidelity polymerase (i.e., Phusion) to reduce the likelihood that mutations will be introduced into products during PCR.
4. Once the final PCR amplification of selected and unselected products is complete, run a small aliquot onto an agarose gel. Confirm that the majority of product present is the band of expected size (*see* **Note 2**).

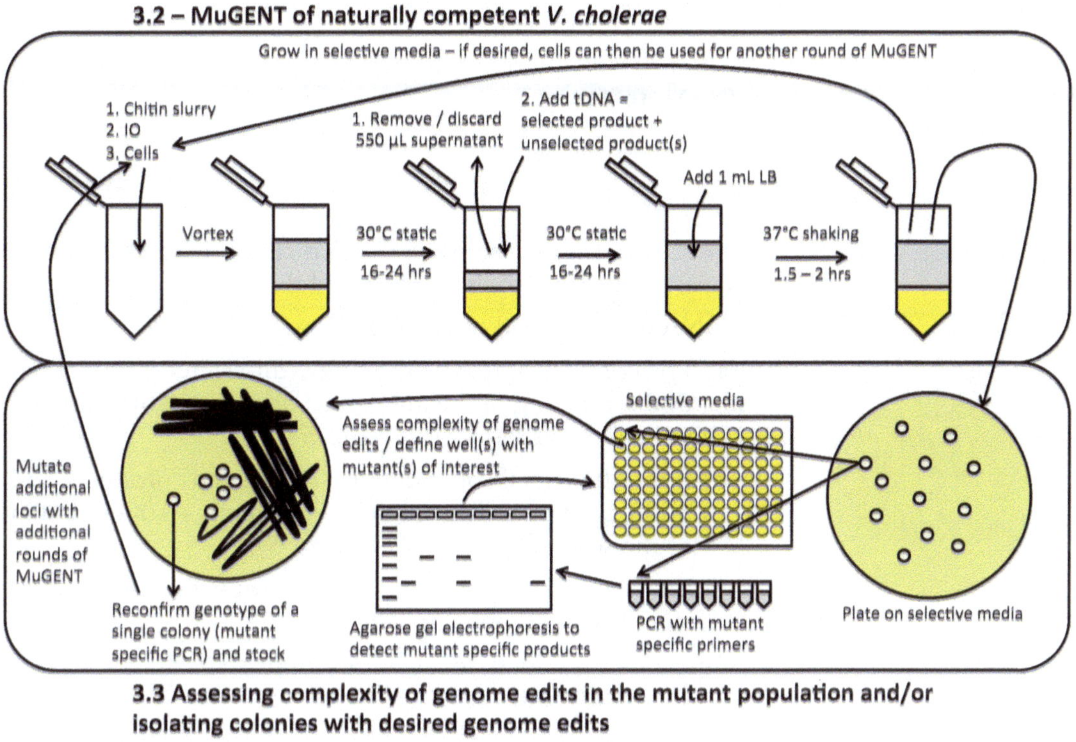

Fig. 2 Detailed diagram outlining the steps for MuGENT and for assessing/selecting edited strains. Selective media = media selective for integration of the selected product

5. Cleanup PCR reactions using a PCR purification kit according to manufacturer's instructions.
6. Next, quantify the concentration of the selected and unselected products on a microvolume spectrophotometer (assess A_{260} and A_{280}).

3.2 MuGENT of Naturally Competent *V. cholerae*

1. Grow a 3 mL culture of the *V. cholerae* strain to be mutated in LB broth in a roller drum at 30 °C overnight (12–18 h) (*see* **Note 3**).
2. The following day, subculture 10–30 µL of this overnight culture (OD_{600} of overnight ≈4–5) into 3 mL of fresh LB broth and grow at 30 °C in a roller drum to an OD_{600} of between 0.8 and 1.5.
3. Spin 1 mL of culture, carefully remove and discard the supernatant, and resuspend the cell pellet with IO to achieve an OD_{600} of 1.0 (i.e., if cells were at an OD_{600} of 0.8, resuspend the cell pellet in 800 µL IO). 100 µL of cells are needed for each cotransformation reaction. If more cells are needed, repeat this step with additional culture.

4. Shake chitin slurry to resuspend settled chitin powder, and transfer 150 μL into a 2 mL microcentrifuge tube for each cotransformation reaction. Add 750 μL of IO to each reaction (*see* **Note 4**).
5. Then, add 100 μL of cells (prepared in **step 3**) to each reaction. Vortex to mix and incubate at 30 °C statically for 16–24 h.
6. Next, without disturbing the settled chitin, remove 550 μL of supernatant from each reaction and discard.
7. tDNA is now added to each reaction:
   (a) For cotransformation of a single unselected product, add 3000–5000 ng of the unselected product + 100–500 ng of the selected product (*see* **Note 5**).
   (b) For MuGENT of multiple unselected products, add 3000 ng of up to five unselected products + 1000–3000 ng of the selected product (*see* **Note 6**).
   (c) Tap the tubes lightly, and invert gently 1–2 times to mix tDNA with competent cells on chitin.
8. Once DNA is added, incubate reactions statically for 16–24 h at 30 °C (*see* **Note 7**).
9. Next, vortex each reaction vigorously for 5–10 s, and add 1 mL of LB broth. Shake reactions at 37 °C for 1.5–2 h to outgrow (*see* **Note 8**). If additional rounds of MuGENT are not needed, proceed to Subheading 3.3.
10. If additional rounds of MuGENT are needed to increase the complexity of unselected genome edits in the mutant population, take 0.5 mL of the outgrown reaction (from **step 9**), and add this to 100 mL of medium that is selective for integration of the selected product [i.e., into 100 mL of LB broth plus kanamycin (50 μg/mL), spectinomycin (200 μg/mL), trimethoprim (10 μg/mL), or streptomycin (100 μg/mL) depending on what the selected product confers resistance to]. Incubate at 37 °C with shaking overnight (12–18 h). The following day, perform this protocol again starting at Subheading 3.2, **step 2** (i.e., subculture 30 μL of this culture into 3 mL of fresh medium).
11. The selected product used in the subsequent round of MuGENT should confer resistance to a different antibiotic to maintain selection at each round of this process (i.e., if in the first round of MuGENT, a Δ*lacZ*::Kan^R marker was used, in the next round, a Δ*lacZ*::Spec^R can be used) (*see* **Note 9**).

3.3 Assessing Complexity of Genome Edits in the Mutant Population and/or Isolating Colonies with Desired Genome Edits

1. Spread plate 100 μL of the neat, 10^{-1}, 10^{-2}, 10^{-3}, and 10^{-4} dilutions of the outgrown reactions (Subheading 3.2, **step 9**) onto media that is selective for your selected product [i.e., onto kanamycin (50 μg/mL), spectinomycin (200 μg/mL), trimethoprim (10 μg/mL), or streptomycin (100 μg/mL) plates depending on what your selected product confers resistance to]. Incubate plates at 30 °C overnight (12–18 h).
2. Next, you will pick and screen single colonies. To assess the complexity of mutations at each round of MuGENT, pick 48 colonies or more. To isolate a colony with a desired combination of genome edits, pick 8–96 colonies depending on the expected frequency of the mutant desired (*see* **Note 10**). To prepare, fill a well of a 96-well plate for each colony to be picked with 200 μL of LB broth with the appropriate antibiotic (i.e., containing the antibiotic that is selective for integration of your selected product) (*see* **Note 11**). Also, fill a PCR tube with 50 μL of sterile H_2O for each colony to be picked. To pick colonies, use sterile pipette tips (or sterile toothpicks) to lift the colony from the plate, and then lightly dab this tip into the selective media in the 96-well plate, and then resuspend the remainder of the colony in the 50 μL of H_2O in the PCR tube (*see* **Note 12**). Once all colonies are picked, incubate the 96-well plate at 37 °C to grow selected colonies.
3. Then, boil the PCR tubes containing the resuspended colonies in a thermal cycler for 5 min. After boiling, vortex reactions. This will serve as template for mutant-specific colony PCR. If performing MuGENT with more than one unselected genome edit, integration of multiple mutations can be assessed via multiplex allele-specific colony PCR (MASC-PCR) [16].
   (a) Primer design for single locus colony PCR—detection oligos for specific mutations should be designed where the F oligo will only amplify the mutant sequence. The R oligo will dictate the size of the product that will be made if the F oligo amplifies the mutant sequence. Products should be designed to be <1000 bp because colony PCR with *Taq* often fails for products larger than this. Generally, mutant-specific oligos are designed to place the mutation within the 3′ end of the oligo sequence because annealing of the 3′ end of an oligo is essential for PCR. Design criteria for primers to distinguish mutant from wild-type sequence for different types of mutations are discussed below:
      (i) For point mutations, this can be accomplished via mismatch amplification mutation assay (MAMA)-PCR [17], where the 3′ most base pair of the detection oligo is specific for the mutant sequence and the second to last bp on the 3′ end is a transition muta-

tion (i.e., the second to last base will mismatch both the wild-type and mutant sequences; if the second to last base is a C, then mutate it to a T in the detection oligo and vice versa. If the second to last base is a G, then mutate it to an A in the detection oligo and vice versa). This promotes enhanced discrimination by allowing for amplification of the mutant sequence while preventing amplification of the wild-type sequence.

(ii) For deletions, the detection oligo can be designed to span the deletion so that the 3–5 3′ most bp are complementary to one side of the deletion, while the remainder of the oligo is complementary to the other side of the mutation. In the example sequence AAAAAAAAAXXXXXXXXXBBBBBB where the desired genome edit will delete the XXXXXXXXX sequence, the detection oligo could be AAAAAAAAABBB. Ensure that the bases BBB in this detection oligo differ from the sequence immediately adjacent to the wild-type sequence (i.e., ensure that the underlined Xs in AAAAAAAAAXXXXXXXXXBBBBBB are different from the Bs in AAAAAAAAABBB). As discussed above for point mutations, ensuring that at least the last two bases on the 3′ end of the oligo mismatch the wild-type sequence will prevent amplification.

(iii) For insertions, the entire oligo can be complementary to the inserted sequence, or the 3′ end of the detection oligo can be within the inserted sequence.

(b) Primer design for MASC-PCR: adhere to all design constraints mentioned above. For detection of multiple genome edits, the R oligos for each locus should be designed to amplify amplicons of different sizes so that products can be easily differentiated on an agarose gel. For MASC-PCR, we generally design detection oligos to simultaneously detect up to seven genome edits with detection amplicons of 150, 200, 300, 400, 550, 700, and/or 950 bp.

4. Use boiled templates to setup 25 μL colony PCR reactions with *Taq* polymerase (*see* **Note 13**):

5 μL 5× *Taq* reaction buffer

0.5 μL Forward primer(s) (30 μM)

0.5 μL Reverse primer(s) (30 μM)

2.5 μL Boiled colony template

16 μL milliQ H_2O

0.5 μL *Taq* polymerase (2000 U/mL)

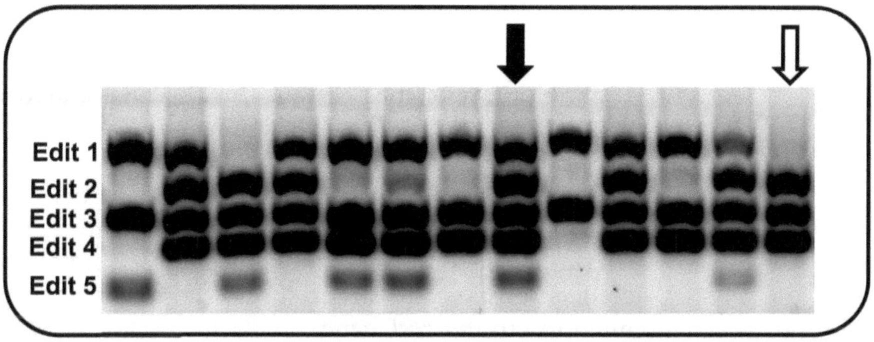

Fig. 3 MASC-PCR for detection of five distinct genome edits. Each lane represents an individual mutant strain, and the presence of a band indicates the presence of a genome edit. The open arrow indicates a strain that received three genome edits (Edits 2, 3, and 4), while the filled arrow indicates a strain that received all five genome edits

Cycling conditions:

95 °C 2 min

23 cycles of 95 °C 40 s, 58 °C 30 s, 72 °C 3 min

72 °C 3 min

5. Run PCR reactions on a 2% agarose gel to assess the presence or absence of genome edits. An example of a MASC-PCR gel is shown in Fig. 3.
6. Retrieve the 96-well plate from the 37 °C incubator, and confirm that colonies have grown by visually assessing turbidity. Determine which wells contain mutants of the desired genotype by consulting your agarose gel. Ideally, you will have two independent wells with the desired genotype. If so, you should perform the subsequent steps on both independently.
7. Next, reconfirm a single colony isolate from the desired wells of the 96-well plate (*see* **Note 14**). To that end, take 20 μL from the well containing the desired mutant, and spot onto a selective plate (i.e., a plate with the antibiotic selective for the selected product). While still wet, perform a quadrant streak for single colonies from the 20 μL spot. Incubate this plate statically at 30 °C overnight (12–18 h) (*see* **Note 15**).
8. Pick a single colony from each plate using a sterile pipette tip as described above, and (1) dab into 3 mL of fresh LB in a culture tube, and (2) resuspend the remainder of the colony in 50 μL of H_2O in a PCR tube.
9. Incubate the culture tubes at 30 °C rolling, and boil the PCR tubes to serve as template for colony PCR.
10. Perform **steps 4** and **5** on the boiled template to reconfirm that all desired genome edits are present.
11. Once confirmed, stock cultures when turbid (can take between 8 and 18 h of growth) by adding 1 mL of culture to 333 μL of 80% glycerol in a cryovial. Freeze at −80 °C.

4 Notes

1. If antibiotic resistance cassettes in the final edited strain are not desired, use auxotrophic strains of *V. cholerae*, and use a selected product that repairs the auxotrophic mutation. Also, it is common to use streptomycin resistant derivatives of *V. cholerae* in lab studies. Thus, for cotransformation, use a streptomycin-sensitive strain, and use a selected product that confers resistance to streptomycin (commonly this is a K43R mutation to the *rpsL* gene).
2. For selected products, even if a faint band of the expected size is obtained (with other bands present), this will suffice, although the total number of transformants obtained during cotransformation may be reduced. For the unselected product, it is critical that the product is predominantly that of the expected size.
3. If a roller drum is not available, cultures can also be grown with shaking at 30 °C in this and all subsequent steps. Common lab strains of *V. cholerae* that have been successfully mutated via this procedure include C6706, A1552, and E7946.
4. Be sure to use chitin powder in this procedure. We have seen that larger chitin fragments and different chitin preparations can result in variable and reduced transformation efficiencies. Pipetting the chitin powder slurry can be difficult, as it will clog most 1 mL pipette tips. To make this process easier, cut the end of a 1 mL pipette off with sterile scissors (ethanol sterilized or autoclaved).
5. Increasing the amount of unselected product can enhance the cotransformation efficiency, while increasing the amount of selected product will increase the number of transformants obtained. 500 ng of selected product can yield between 10^4 and 10^6 transformants.
6. If fewer than five unselected products are being added, 5000 ng of each product can be added as long as the total amount of unselected DNA added does not exceed 15,000 ng (i.e., for integration of three unselected products, 5000 ng of each can be added to the reaction).
7. If in a rush, DNA can also be incubated on cells for 5 h prior to continuing with the protocol. In most cases this will yield cotransformation efficiencies almost as high as when cells are incubated with DNA overnight.
8. This step helps to "resolve" and segregate mutations as well as break up any clumps of bacteria to ensure that when plated, each colony is clonal. Longer outgrowth is better as long as integration of the unselected mutation does not decrease the growth rate of the cell.

9. Using selected markers in subsequent rounds of MuGENT that replace the antibiotic marker from a previous round is beneficial because this allows the use of the first marker in later rounds of MuGENT if desired (i.e., using Δ*lacZ*::Kan^R marker in the first round of MuGENT and then Δ*lacZ*::Spec^R in the next round allows the use of Δ*lacZ*::Kan^R in subsequent rounds).
10. For cotransformation of a single unselected product, efficiencies can reach 50%, so screening 8–16 colonies will usually suffice to obtain a mutant with the desired genome edit. To obtain a strain that has 3–4 unselected genome edits, the expected efficiency would be the product of the efficiency of each product independently. If we assume a cotransformation efficiency of 40% for each product, then we would expect an efficiency of 2.5% for a mutant containing all four genome edits (i.e., $0.4 \times 0.4 \times 0.4 \times 0.4$). Thus, screening 96 colonies may result in two of the desired mutant.
11. If mutants are being screened solely to assess the complexity of genome edits, then picking colonies into a 96-well plate for growth is not required.
12. Be sure to keep track of which PCR tube corresponds to which well of the 96-well plate. Once the genotype of the strain is determined by colony PCR, you will need to pick that strain from the 96-well plate.
13. If performing MASC-PCR, the total amount of oligo added to the PCR reaction remains the same [i.e., for detection of five genome edits, 0.1 μL of each F and R detection oligo (30 μM) would be used instead of 0.5 μL (30 μM) as indicated for one genome edit]. The extension time of 3 min for the 72 °C step during PCR cycling is to reduce the preferential amplification bias of smaller amplicons when performing MASC-PCR. If only a single locus is being screened for, the extension time can be reduced to 1 min.
14. Despite outgrowth (Subheading 3.2, **step 9**), sometimes clumps of cells remain when plating cotransformation reactions, which results in a heterogeneous colony. Therefore, it is prudent to reconfirm a single colony for desired mutants as indicated.
15. To save the remainder of the colonies picked on the 96-well plate, add 66 μL of 80% glycerol to each well of the plate (i.e., final concentration of ~20%). Thoroughly mix by pipetting and freeze at −80 °C.

References

1. Lorenz MG, Wackernagel W (1994) Bacterial gene transfer by natural genetic transformation in the environment. Microbiol Rev 58(3):563–602 Epub 1994/09/01
2. Meibom KL, Blokesch M, Dolganov NA, Wu CY, Schoolnik GK (2005) Chitin induces natural competence in *Vibrio cholerae*. Science 310(5755):1824–1827. Epub 2005/12/17. https://doi.org/10.1126/science.1120096
3. Chen Y, Dai J, Morris JG Jr, Johnson JA (2010) Genetic analysis of the capsule polysaccharide (K antigen) and exopolysaccharide genes in pandemic Vibrio parahaemolyticus O3:K6. BMC Microbiol 10:274. Epub 2010/11/04. https://doi.org/10.1186/1471-2180-10-274
4. Gulig PA, Tucker MS, Thiaville PC, Joseph JL, Brown RN (2009) USER friendly cloning coupled with chitin-based natural transformation enables rapid mutagenesis of Vibrio vulnificus. Appl Environ Microbiol 75(15):4936–4949. Epub 2009/06/09. https://doi.org/10.1128/AEM.02564-08
5. Pollack-Berti A, Wollenberg MS, Ruby EG (2010) Natural transformation of Vibrio fischeri requires tfoX and tfoY. Environ Microbiol 12(8):2302–2311. Epub 2010/08/01. https://doi.org/10.1111/j.1462-2920.2010.02250.x
6. Blokesch M (2012) TransFLP—a method to genetically modify *Vibrio cholerae* based on natural transformation and FLP-recombination. J Vis Exp 68. Epub 2012/10/25. https://doi.org/10.3791/3761
7. Dalia AB, Lazinski DW, Camilli A (2014) Identification of a membrane-bound transcriptional regulator that links chitin and natural competence in *Vibrio cholerae*. MBio 5(1):e01028-13. https://doi.org/10.1128/mBio.01028-13
8. Dalia AB, Seed KD, Calderwood SB, Camilli A (2015) A globally distributed mobile genetic element inhibits natural transformation of *Vibrio cholerae*. Proc Natl Acad Sci U S A 112(33):10485–10490. https://doi.org/10.1073/pnas.1509097112
9. Baba T, Ara T, Hasegawa M, Takai Y, Okumura Y, Baba M et al (2006) Construction of *Escherichia coli* K-12 in-frame, single-gene knockout mutants: the Keio collection. Mol Syst Biol 2:2006 0008. Epub 2006/06/02. https://doi.org/10.1038/msb4100050
10. Zhang Y, Buchholz F, Muyrers JP, Stewart AF (1998) A new logic for DNA engineering using recombination in *Escherichia coli*. Nat Genet 20(2):123–128. https://doi.org/10.1038/2417
11. Sung CK, Li H, Claverys JP, Morrison DA (2001) An rpsL cassette, janus, for gene replacement through negative selection in Streptococcus pneumoniae. Appl Environ Microbiol 67(11):5190–5196. https://doi.org/10.1128/AEM.67.11.5190-5196.2001
12. Erickson RJ, Copeland JC (1973) Congression of unlinked markers and genetic mapping in the transformation of Bacillus subtilis 168. Genetics 73(1):13–21 Epub 1973/01/01
13. Dalia AB, McDonough E, Camilli A (2014) Multiplex genome editing by natural transformation. Proc Natl Acad Sci U S A 111(24):8937–8942. https://doi.org/10.1073/pnas.1406478111
14. Horton RM, Cai ZL, Ho SN, Pease LR (1990) Gene splicing by overlap extension: tailor-made genes using the polymerase chain reaction. Biotechniques 8(5):528–535
15. Horton RM, Hunt HD, Ho SN, Pullen JK, Pease LR (1989) Engineering hybrid genes without the use of restriction enzymes: gene splicing by overlap extension. Gene 77(1):61–68. Epub 1989/04/15. https://doi.org/10.1016/0378-1119(89)90359-4
16. Wang HH, Isaacs FJ, Carr PA, Sun ZZ, Xu G, Forest CR et al (2009) Programming cells by multiplex genome engineering and accelerated evolution. Nature 460(7257):894–898. Epub 2009/07/28. https://doi.org/10.1038/nature08187
17. Cha RS, Zarbl H, Keohavong P, Thilly WG (1992) Mismatch amplification mutation assay (MAMA): application to the c-H-ras gene. PCR Methods Appl 2(1):14–20

Chapter 7

Chromatin Immunoprecipitation

Julio C. Ayala, Jorge A. Benitez, and Anisia J. Silva

Abstract

Chromatin immunoprecipitation (ChIP) measures the physical association between a protein and DNA in the cell. In combination with next-generation sequencing, the technique enables the identification of DNA targets for the corresponding protein across an entire genome. Here we describe the immunoprecipitation of *Vibrio cholerae* DNA bound to the histone-like nucleoid structuring protein (H-NS) tagged with the Flag epitope. The quality of the DNA obtained in this protocol is suitable for next-generation sequencing. The procedure described herein can be readily adapted to other bacteria and DNA-binding proteins.

Key words Transcription, DNA-binding proteins, Immunoprecipitation, Chromatin immunoprecipitation, Next-generation sequencing, *Vibrio cholerae*

1 Introduction

Bacteria are notorious for their capacity to respond to environmental stresses and adapt to changes in the physical-chemical characteristics of the extracellular milieu. This adaptability allows them to effectively compete for nutrients, persist within natural reservoirs, establish commensal or symbiotic relationships with other organisms, or cause disease in plants, animals, and humans. Adaptations to environmental changes require a continuous readjustment of gene expression patterns. A major goal of molecular biology is to determine the mechanism by which genes are activated or turned off in response to environmental cues. The interaction between transcription factors and promoters is commonly the final step of many signal transduction pathways. In addition to transcription regulation, DNA-binding proteins play essential roles in the organization of the nucleoid, genetic recombination, the modification, repair and replication of DNA, and chromosome segregation. Hence, the measurement of the association between specific proteins and DNA sequences is a core molecular biology protocol. Electrophoresis mobility shift assays, wherein binding of a protein

Aleksandra E. Sikora (ed.), *Vibrio Cholerae: Methods and Protocols*, Methods in Molecular Biology, vol. 1839,
https://doi.org/10.1007/978-1-4939-8685-9_7,

to a DNA fragment retards its electrophoretic migration in a polyacrylamide gel, has been extensively used to measure protein-DNA interactions in vitro. This technique, though extremely useful, may not accurately reflect the interaction between proteins and the native chromatin in the cell. Chromatin immunoprecipitation (ChIP) provides a method to quantitate the physical association between a protein and its DNA target in vivo [1, 2]. In this method, the protein under study can be tagged with an epitope so that it can be immunoprecipitated with a specific monoclonal antibody (mAb). The gene encoding the epitope-tagged protein is expressed in the host bacterium and the protein is cross-linked to DNA. Next, the cells are lysed and chromosomal DNA is broken down by sonication to fragments ranging in size from 200 to 1000 bp. Protein-DNA complexes are immunoprecipitated and washed and the cross-linking is reversed. Finally, the immunoprecipitated DNA is purified and the quantity of protein-bound target DNA measured by quantitative real-time PCR (qRT-PCR) [1, 2]. A flow diagram of the complete procedure is shown in Fig. 1. If the DNA target is unknown, or if the goal is to identify all DNA targets of a particular protein, ChIP can be combined with next-generation sequencing to reveal the binding profile of the corresponding protein across the entire genome [3]. In this protocol, we describe the immunoprecipitation of *Vibrio cholerae* DNA bound to the histone-like nucleoid structuring protein (H-NS) tagged with the Flag epitope [4–7]. The protocol can be adapted to other bacteria and DNA-binding proteins and yields DNA of a quality suitable for targeted qRT-PCR and next-generation sequencing. The protocol assumes that a bacterium expressing the epitope-tagged DNA-binding protein of interest is available.

2 Materials

2.1 Bacterial Culture and Cross-linking

1. *Vibrio cholerae* strain C7258H-NS-FLAG [6] (*see* **Note 1**).
2. Bacteriological media: Luria-Bertani (LB) agar and tryptic soy broth (TSB) (*see* **Note 2**).
3. Formaldehyde 37%.
4. Glycine 2.5 M stock solution (*see* **Note 3**). Store at 4 °C (*see* **Note 4**).
5. Phosphate-buffered saline pH 7.4: dissolve 8 g NaCl, 0.2 g KCl, 1.44 g Na_2HPO_4, and 0.24 g KH_2PO_4 in 800 mL of water. Adjust the pH to 7.4 with HCl, complete volume to 1 L with water, and sterilize by filtration.
6. Protease inhibitor cocktail (PIC) stock solution (Roche Applied Sciences catalog # 11836153001): Dissolve one tablet

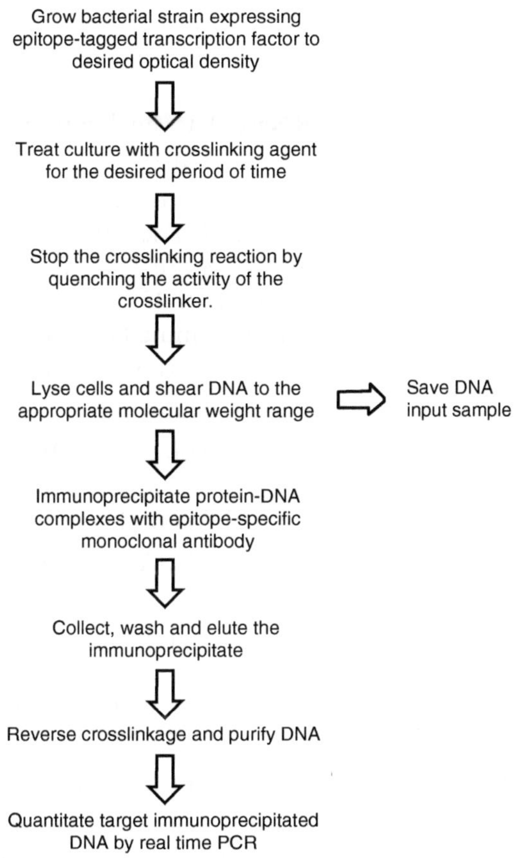

Fig. 1 The ChiP workflow. For a specific protein, two ChIP assays are conducted in parallel: one using a monoclonal antibody that reacts with the epitope attached to the protein (problem ChIP) and a second assay using an unrelated monoclonal antibody that does not cross-react with any bacterial host protein (mock ChIP). Prior to immunoprecipitation, a sample is taken from each ChIP assay to quantitate the starting amount of target DNA (input sample). ChIP is expressed as the percentage of target DNA present in the immunoprecipitate relative to the input sample. ChIP can also be expressed as the amount of this DNA present in the problem ChIP normalized by the quantity detected in the mock ChIP (relative ChIP)

in 0.4 mL sterile water to obtain a 25× strength solution. Store at −20 °C.

7. Phenylmethylsulfonyl fluoride (PMSF) 100× strength stock solution: 100 mM PMSF in absolute ethanol. Store at −20 °C.

2.2 Cell Lysis and Sonication

1. RNase A stock solution: dissolve 10 mg of RNase A in 1 mL sterile water.
2. Ready-lyse lysozyme (Epicentre Biotechnologies catalog # R1802M).

3. Lysis buffer: 10 mM Tris–HCl pH 7.5, 50 mM NaCl, 1 mM ethylenediaminetetraacetic acid (EDTA). Store at 4 °C. Immediately before use, add Ready-lyse lysozyme and RNase A to final concentrations of 14.5 kU/mL and 20 μg/mL, respectively.
4. Double strength (2×) immunoprecipitation (IP) buffer: 200 mM Tris–HCl pH 7.5, 600 mM NaCl, and 4% triton X-100. Store at 4 °C. To 10 mL of the above solution, add 0.08 mL of 25× PIC stock solution and 0.2 mL of PMSF stock solution immediately before use. Prepare IP buffer by diluting 1:2 the 2× IP solutions in sterile water.
5. Phenol/chloroform.
6. Misonix sonicator S-4000 with the microtip or equivalent.

2.3 Immuno-precipitation

1. Anti-FLAG monoclonal antibody. Store at −20 °C.
2. Mouse (G3A1) mAb IgG1 Isotype Control. Stored at −20 °C.
3. End-over-end rotator.

2.4 Immuno-precipitate Collection, Washing, and Elution

1. Salmon Sperm DNA/Protein G Agarose (Novus Biologicals catalog # KC-129).
2. Costar Spin-X centrifuge tube 0.22 μm filters.
3. Wash buffer A: 100 mM Tris–HCl pH 7.5, 2% triton X-100, 250 mM LiCl (*see* **Note 5**).
4. Wash buffer B: 100 mM Tris–HCl pH 7.5, 2% triton X-100, 600 mM NaCl.
5. Wash buffer C: 100 mM Tris–HCl pH 7.5, 2% triton X-100, 300 mM NaCl.
6. TE buffer: 10 mM Tris–HCl pH 8.0, 1 mM EDTA.
7. Elution buffer: 10 mM Tris–HCl pH 8.0, 1 mM EDTA, 1% sodium dodecyl sulfate (SDS).
8. Proteinase K stock solution: 10 mg/mL in sterile water.

2.5 Reversal of Cross-linking, DNA Purification, and Real-Time PCR

1. MiniElute PCR purification kit (Qiagen catalog # 28006).
2. iTaq™ Universal SYBR Green Super mix (BioRad catalog # 172-512).
3. CFX96 Touch real-time PCR detection system (BioRad) or equivalent.
4. CFX96 Touch real-time PCR detection system accessories (i.e., 96-well PCR plates, optical tape).

3 Methods

3.1 Bacterial Culture and Cross-linking

1. Scratch a few crystals from a frozen stock of the strain that expresses the epitope-tagged DNA-binding protein (i.e., C7258H-NS-FLAG), and streak on LB agar to obtain isolated colonies. Incubate overnight at 37 °C.
2. Inoculate a well-isolated colony into 5 mL of TSB medium in a 50 mL Falcon centrifuge tube. For conducting this assay in triplicate (recommended), inoculate three Falcon tubes containing 5 mL of TSB. Set orbital shaker at 225 rpm, and incubate 16 h at 37 °C with agitation.
3. Transfer 0.5 mL of the overnight-grown culture(s) to 50 mL of fresh TSB medium in a 250-mL culture bottle, and incubate at 37 °C with agitation until the culture reaches the desired optical density at 600 nm (OD_{600}).
4. Add 1.1 mL of 37% formaldehyde to 40 mL of culture to crosslink proteins and DNA. Incubate at 30 °C with agitation for 10 min (*see* **Note 6**).
5. Distribute 10 mL of the formaldehyde-treated cell suspension into four 28 mL sterile Oak-Ridge Nalgene centrifuge tubes.
6. Add 1 mL of glycine stock solution to each tube to stop the cross-linking reaction, and incubate at 4 °C in an end-over-end rotator for 30 min.
7. Centrifuge tubes at 4649 × *g* and 4 °C for 10 min. Discard the supernatant, and wash the pellet twice by resuspending cells in 10 mL of ice-cold PBS solution and centrifugation.
8. Reconstitute each pellet in 1 mL of ice-cold PBS supplemented with 40 μL of PIC and 10 μL of PMSF stock solution.
9. Aliquot each cell suspension into 1.5 mL nuclease-free microfuge tubes. Aliquots should contain approximately 3×10^8 colony-forming units (CFU) of the strain used (*see* **Note 7**).
10. Centrifuge tubes at 16,000 × *g* for 5 min at 4 °C.
11. Remove the supernatant carefully with a pipettor without disturbing the pellet.
12. The cell pellet can be stored at −80 °C for several months.

3.2 Cell Lysis and Sonication

1. Reconstitute cell pellet from previous step in 0.5 mL of lysis buffer, and incubate at 37 °C for 30 min.
2. Add 0.5 mL of 2× IP buffer.
3. Transfer cell suspension to a 15 mL Falcon centrifuge tube, and sonicate on ice at 25% amplitude (power ≈ 7 Watt) using

the Misonix sonicator S-4000 with the microtip. Deliver repeated cycles of 15 s separated by 2 min on ice (*see* **Note 8**).

4. Transfer the sheared DNA solution to a 1.5 mL microfuge tube, and centrifuge at 16,000 × *g* 10 min at 4 °C.
5. Transfer 0.1 mL aliquots of the supernatant to clean 1.5 mL microfuge tubes, and store at −80 °C until use. Check average molecular weight of the sheared DNA by agarose gel electrophoresis (*see* **Note 9**).

3.3 Immunoprecipitation

1. Mark two vials containing 0.1 mL of sheared material as problem ChIP and mock ChIP, respectively.
2. Complete the volume in each vial to 1 mL by adding 0.9 mL of IP buffer.
3. Take a sample of 10 μL from each tube and store at −20 °C. These samples (input samples) are used to quantitate the starting amount of target DNA (*see* **Note 10**).
4. Add 8 μL (8 μg) of anti-FLAG mAb to the problem ChIP tube (*see* **Note 11**).
5. Add 2.2 μL (8 μg) of the mouse (G3A1) mAb IgG1 isotype control to the mock ChIP tube (*see* **Note 12**).
6. Close and seal the tubes with parafilm and rotate overnight at 4 °C. From this step on, the problem and mock ChIP samples are subject to the same treatment.

3.4 Immunoprecipitate Collection, Washing, and Elution

1. Add 50 μL of Salmon Sperm DNA/Protein G Agarose to each immunoprecipitate (*see* **Note 13**).
2. Close and seal the tubes with parafilm and rotate 1 h at 4 °C.
3. Centrifuge at 1100 × *g* for 30 s at room temperature.
4. Discard supernatant carefully to avoid removing the agarose beads. For instance, remove the bulk of the supernatant with a 1 mL pipettor, and then aspirate the remaining with a smaller volume pipettor.
5. Suspend agarose beads in 1 mL of wash buffer A by tapping the tube. Centrifuge at 1100 × *g* 30 s, and remove supernatant as in previous step.
6. Suspend agarose beads in 0.8 mL of wash buffer A, and transfer to a Spin-X centrifuge tube filter (*see* **Note 14**). Rotate the Spin-X centrifuge tube filter for 2–3 min at 4 °C and centrifuge 1 min at 4 °C (*see* **Note 15**).
7. Discard the flow through and add 0.8 mL of wash buffer A. Rotate the Spin-X centrifuge tube filter for 2–3 min at 4 °C, and centrifuge 1 min at 4 °C.
8. Repeat **step 7** three times with 0.8 mL of wash buffer B.

9. Repeat **step 7** three times with 0.8 mL of wash buffer C.
10. Repeat **step 7** three times with 0.8 mL of TE buffer.
11. Discard the last flow through, resuspend the agarose bead slurry in 0.25 mL of elution buffer, and transfer the full content to a clean microfuge tube. To maximize recovery, follow the suggestion indicated in **Note 14** but using smaller volumes.
12. Add 250 μL of elution buffer to the corresponding input samples taken in **step 3** of Subheading 3.3, and mix gently by tapping the tubes.
13. Incubate tubes from **steps 11** and **12** at 65 °C 30 min with occasional shaking.
14. Centrifuge at 3100 × *g* for 30 s and collect the supernatant. Recover as much supernatant as possible, and transfer to a clean microfuge tube.

3.5 Reversal of Cross-linking, DNA Purification, and Real-Time PCR

1. Add 10 μL of 5 M NaCl, and incubate at 65 °C for 4 h (or overnight) to reverse cross-linking. At this stage samples can be stored at −20 °C if necessary.
2. Add 5 μL of 0.5M EDTA, 10 μL 1 M Tris–HCl pH 7.0, and 2 μL proteinase K stock solution to each tube, and incubate 1 h at 45 °C.
3. Purify DNA from input and immunoprecipitated samples using the MiniElute PCR purification kit following the manufactures protocol.
4. Elute DNA with 0.1–0.2 mL of sterile water.
5. Mix 12.5 μL of iTaq Universal SYBR Green Super mix, 1 μL of each PCR primer (10 ρmoles) (*see* **Note 16**), 8.5 μL sterile water, and 2 μL of input or immunoprecipitated samples. Set up the above reactions in duplicate (two wells per sample).
6. The reaction mixtures are cycled in the in CFX96 real-time PCR detection system (*see* **Note 17**).
7. The amount of immunoprecipitated (IP) target DNA is calculated as a percentage of the amount of target DNA present in the input sample using the formula $\% \text{ IP} = 2^{(\text{Ct input} - \text{Ct IP})}$ in which Ct is the fractional threshold cycle determined for the problem and mock ChiP reactions and their corresponding input samples [8, 9] (*see* **Note 18**). An example of the results obtained following the above protocol is shown in Table 1 (*see* **Note 19**).
8. A standard PCR protocol and agarose gel electrophoresis can be used for the qualitative detection of target DNA in immunoprecipitated samples.

Table 1
ChIP results for H-NS binding to the *V. cholerae tcpA* promoter

Gene	% Immunoprecipitation $2^{(\text{Ct input} - \text{Ct immunoprecipitated})}$		Relative IP (problem ChIP/mock ChIP)
	Problem ChIP	mock ChIP	
tcpA	0.18 224 ± 0.01504	0.00023 ± 0.00005	808.95 ± 121.55
rpsM	0.00107 ± 0.00006	0.00021 ± 0.00003	5.27 ± 0.95

4 Notes

1. This strain expresses the histone-like nucleoid structuring protein (H-NS) tagged with the Flag epitope from native transcription and translation initiation signals. The *hns*-Flag allele was constructed using standard recombinant DNA techniques and the pFLAG-CTC™ expression vector available from Sigma-Aldrich (catalog # E8408). The allele was integrated into the *hns* locus by homologous recombination, therefore disrupting the native *hns* gene. It is possible to express an epitope-tagged DNA-binding protein on a plasmid vector. In this case we recommend (i) the use of a low-copy-number vector to minimize nonspecific binding resulting from overexpression of the DNA-binding protein and (ii) introducing the plasmid in a deletion mutant lacking the native chromosomally encoded protein to avoid competition with the epitope-tagged DNA-binding protein for binding to DNA. We recommend the preparation of frozen stocks of the strain to be used. This can be achieved by collecting biomass from a heavily streaked agar plate into 5 mL of tryptic soy broth containing 25% glycerol.
2. LB and TSB media are commercially available from several vendors. These media can be replaced as needed or supplemented with antibiotics to meet the specific requirements of other bacteria.
3. Prepare all solutions with analytical grade reagents of the highest purity dissolved in sterile ultrapure 18 MΩ deionized nuclease-free water. We recommend handling solutions inside a biosafety cabinet and transfer volumes using aerosol-resistant filter pipette tips.
4. In case of precipitation, warm up before use and dissolve glycine by vortexing.
5. We recommend adding 0.1 volumes of a 2.5 M LiCl stock solution filtered through a 0.22 μm syringe filter and autoclaved before use.

6. Cross-linking time and temperature must be strictly controlled. Insufficient or excessive cross-linking can result in DNA loss or high background. Cross-linking can influence the efficiency of DNA shearing and antigen immunoprecipitation. This step may be optimized by testing different times (i.e., 10, 15, 20 min). Optimization can be particularly helpful for DNA-binding proteins that are expressed at a low level or interact weakly with their target DNA.
7. We recommend constructing a standard curve of OD_{600} versus CFU for each strain and medium to be used. Then, calculate the aliquot volume from the OD_{600} reading taken in **step 3** of Subheading 3.1.
8. Typically, 16–18 cycles are required. However, the number of cycles required to obtain a sheared DNA of suitable size should be determined by trial for the sonicator model available, specific bacterial strain, and growth conditions as described in **Note 9**.
9. To determine DNA shearing efficiency, add 4 μL of 5 M NaCl to 100 μL of sheared DNA material, and incubate 4 h at 65 °C. Add 20 μL of water and 80 μL of phenol/chloroform (25:1 v/v). Vortex thoroughly and centrifuge at 16,000 × *g* for 3 min. Transfer the aqueous (upper) phase to a new microfuge tube, and add 1/10 volume of 3 M sodium acetate (pH 6.0) and one volume of isopropanol. Mix gently and incubate at least 20 min at −20 °C. Centrifuge at 16,000 × *g* for 6 min at room temperature and remove supernatant. Wash the pellet with 200 μL of ethanol 80%, centrifuge, and remove supernatant again. Air-dry the pellet at 37 °C. Reconstitute the DNA pellet in 60 μL sterile water, and load 20 μL on a 1% agarose gel alongside with 100 bp ladder molecular weight markers. Shearing should result in a smear of DNA between 200 and 1000 bp.
10. This sample contains 1% of the starting DNA material. We usually freeze this sample at −20 °C, and process along with other samples the next day.
11. The specificity of the mAb used to immunoprecipitate DNA fragments bound by a DNA-binding protein is critical to success. When using a mAb recognizing a different epitope tag or the whole DNA-binding protein of interest, confirm with the vendor that it has been validated for ChIP.
12. Other commercially available mAbs can be used to conduct the mock ChIP as long as they are of the same IgG subclass used for the problem ChIP, and do not cross-react with proteins in the bacterial lysate.
13. Vortex Protein G agarose briefly prior to pipetting to ensure the same amount of beads is transferred to each tube.

14. First add 400 μL of wash buffer A, and suspend the beads by pipetting up and down; transfer the beads to the column tube filter (keep pipette with tip). Use a second pipette to add an additional 400 μL of buffer A to the Eppendorf tube (do not touch the tube). With the first pipette, recover the beads remaining in the walls of the tube, and transfer them to the column.
15. Typically, the washing steps can be done in an ice box by placing the Spin-X columns horizontally above the ice. Centrifuge at 1500 × *g* for 1 min, and discard the flow through in the reservoir tube with a 1 mL aerosol barrier filter tip. Check that the centrifugation step filters all the liquid through the column. If there is still liquid inside the column, centrifuge again, but do not increase the centrifugation speed since it can damage the agarose beads. Then, try to remove all the foam created by triton X-100 in the reservoir tube. Proceed this way in all subsequent washing steps.
16. The PCR products of the genes analyzed must be of similar length (i.e., 180–220 bp) to avoid differences in fluorescence intensity. All primer pairs must have the same isothermal melting points and PCR amplification efficiencies (at least 90% of efficiency). To determine the PCR amplification efficiency of the primers, set up three PCR reactions in duplicate containing genomic DNA diluted 1/10, 1/100, and 1/1000. If you are using the CFX96 Touch Real time PCR Detection System, proceed as described below. Once the reaction is complete go to the "View/Save Data" tab in the CFX96 menu. Enter the appropriate data for the standard curve (i.e., plate position, quantity). Then, go to "PCR standard curve," and collect the results of PCR efficiency (>90%) and correlation coefficient (>0.9). The primer pairs should also yield a single peak in the melting curve analysis module of the real-time PCR machine. Reevaluate your primers if you switch to another real-time PCR machine or model.
17. Thermal cycling parameters should be adapted to the length, base composition, and sequence of the target DNA fragment being amplified.
18. The results can be represented as the percentage of IP target DNA relative to the amount of this DNA present in the input sample. An alternative way to represent the data is to divide the percentage of IP DNA in the problem ChIP by the percentage IP DNA in the mock ChIP (relative IP). We recommend amplifying a DNA fragment that does not interact with the protein under study as a negative control. The amount of this DNA in the immunoprecipitated samples should be negligible.

19. This table shows ChIP data for the binding of *V. cholerae* H-NS to the *tcpA* promoter. H-NS has been shown to bind to this promoter in vitro and silence its transcription [10, 11]. A sequence located within the housekeeping *rpsM* gene encoding the 30S ribosomal subunit protein S13 was used as a negative control. The Ct values for each real-time PCR reaction were collected using the machine default threshold of ten standard deviations above baseline fluorescence. The primers used to amplify the *tcpA* promoter 5′-GTTCATAAT TTCGATCTCCACTCCG and 5′-GTTAACCACACAAA GTCACCTGCAA; the primers used to amplify *rpsM* sequences were 5′-GCAACTGCGTGATGGTGTAGCTAA and 5′-GCTTGATCGGCTTACGCGGACC [6].

Acknowledgments

This work was supported by Public Service Research Grants and a Ruth L. Kirschstein National Research Service Award from the National Institute of Allergy and Infectious Disease, National Institutes of Health, Bethesda, Maryland, USA.

References

1. Kuo MH, Allis CD (1999) In vivo cross-linking and immunoprecipitation for studying dynamic protein:DNA associations in a chromatin environment. Methods 19:425–433
2. Collas P (2010) The current state of chromatin immunoprecipitation. Mol Biotechnol 45:87–100
3. Valouev A, Johnson DS, Sundquist A et al (2008) Genome-wide analysis of transcription factor binding sites based on Chip-Seq data. Nat Methods 5:829–834
4. Ayala JC, Wang H, Benitez JA et al (2015) RNA-Seq analysis and whole genome DNA-binding profile of the histone-like nucleoid structuring protein (H-NS). Genomics Data 5:147–150
5. Ayala JC, Wang H, Silva AJ et al (2015) Repression by H-NS of genes required for the biosynthesis of the *Vibrio cholerae* biofilm matrix is modulated by the second messenger cyclic diguanylic acid. Mol Microbiol 97:630–645
6. Wang H, Ayala JC, Benitez JA et al (2012) Interaction of the histone-like nucleoid structuring protein and the general stress response regulator RpoS at *Vibrio cholerae* promoters that regulate motility and hemagglutinin/protease expression. J Bacteriol 194:1205–1215
7. Wang H, Ayala JC, Silva AJ et al (2012) The histone-like nucleoid structuring protein (H-NS) is a repressor of *Vibrio cholerae* exopolysaccharide biosynthesis (vps) genes. Appl Environ Microbiol 78:2482–2488
8. Livak KJ, Schmittgen TD (2001) Analysis of relative gene expression data using real-time quantitative PCR and the $2^{(\text{-delta delta Ct})}$ method. Methods 25:402–408
9. Yuan JS, Reed A, Chen F et al (2006) Statistical analysis of real-time PCR data. BMC Bioinformatics 7:85. https://doi.org/10.1186/1471-2105-7-85
10. Nye MB, Pfau JD, Skorupski K et al (2000) *Vibrio cholerae* H-NS silences virulence gene expression at multiple steps in the *toxR* regulatory cascade. J Bacteriol 182:4295–4303
11. Wang H, Ayala JC, Benitez JA et al (2015) RNA-Seq analysis identifies new genes regulated by the histone-like nucleoid structuring protein (H-NS) affecting *Vibrio cholerae* virulence, stress response and chemotaxis. PLoS One 10:e0118295. https://doi.org/10.1371/journal.pone.0118295

Chapter 8

Fly Models of *Vibrio cholerae* Infection and Colonization

Alexandra E. Purdy

Abstract

Studies of *Vibrio cholerae* pathogenesis in the context of novel eukaryotic model systems have expanded our understanding of genes that underlie *V. cholerae* interactions with humans, as well as host organisms in the environment. These model systems have also helped uncover new functions for many gene products, revealing previously unknown virulence mechanisms. The *Drosophila* model for *V. cholerae* infection is a powerful tool for discovering new genetic pathways that govern bacterial physiology and colonization in the arthropod gastrointestinal tract. Assays to measure both virulence and colonization have been established and are easily adopted in labs unfamiliar with *Drosophila* work. Experiments to compare survival of flies colonized with different bacterial mutants are simple to perform and can be completed in less than a week, allowing colonization to be quantified and localized easily. The availability of molecular and genetic tools for the fly enables further exploration of host factors that restrict *V. cholerae* colonization and invasive infection. Based on the *Drosophila* system, a house fly (*Musca domestica*) model of *V. cholerae* colonization has also been developed. The new house fly model may prove a useful tool for examining *V. cholerae* infection dynamics in the context of a host carrying a complex microbial community, with a fundamentally different ecology that may increase its chances of acting as a vector for cholera disease.

Key words *Drosophila melanogaster*, Fruit fly, *Vibrio cholerae*, Virulence, Colonization, Survival, Confocal microscopy, House fly, *Musca domestica*

1 Introduction

Across the fields of genetics, neurobiology, and development, numerous discoveries over the past 100 years have been made possible by the use of *Drosophila* as a model organism. Recently, microbiologists and immunologists have found that *Drosophila* is a powerful tool for exploring the genetic, nutritional, and molecular aspects of interactions between a host, its commensal microbiota, and invading pathogens [1–3]. *Drosophila* are convenient, genetically tractable host organisms, and a wide variety of molecular tools can be applied to define their response to infection. At the same time, pathogenesis can be studied from the bacterial perspective with simple, inexpensive assays that measure both colonization and virulence. *Musca domestica* (house flies) are relatives of fruit flies

Aleksandra E. Sikora (ed.), *Vibrio Cholerae: Methods and Protocols*, Methods in Molecular Biology, vol. 1839, https://doi.org/10.1007/978-1-4939-8685-9_8, © Springer Science+Business Media, LLC, part of Springer Nature 2018

that often live in close association with people, acting as vectors for a number of human pathogens [4, 5]. Although genetic tools are not yet available in *Musca*, these flies may be ecologically relevant hosts in which bacterial mechanisms of infection can be studied in a relatively natural setting. Factors that control colonization, such as nutritional status, the presence of commensal microbiota, and specific genes carried by the pathogen, can be examined in house flies using assays similar to those developed in *Drosophila*. With the advent of genomic tools, *Musca* is an attractive model for interrogating interactions between a pathogen and the commensal microbial community harbored in the fly's gastrointestinal tract.

A *Drosophila* model of *V. cholerae* infection developed in Paula Watnick's laboratory (located first at Tufts-New England Medical Center and later at Boston Children's Hospital) has revealed surprising molecular exchanges that take place during infection [6, 7]. The model has provided insight into novel mechanisms through which cholera toxin may alter the barrier function of the gastrointestinal tract [8, 9] and suggested a role for *Vibrio* polysaccharide-dependent biofilms in colonization of a specific region of the fly alimentary canal [10]. In addition to uncovering new functions for known *Vibrio* virulence factors, the *Drosophila* system has emerged as a powerful platform for discovery. Recently, a genetic screen for *V. cholerae* virulence factors in *Drosophila* revealed a novel signaling pathway that modulates host metabolism [11]. The link between *V. cholerae* metabolism and fly physiology was further strengthened in a study that related the locations of ribosome-encoding loci on the two *V. cholerae* chromosomes to in vitro growth rates and colonization of *Drosophila* [12]. Thus far, the *V. cholerae–Drosophila* system has allowed researchers to dissect physiological mechanisms that contribute to pathogenesis in an entirely new context.

Paula Watnick and her colleagues have developed assays that can independently assess colonization and virulence during *V. cholerae* infection of *Drosophila*. In their first publication describing the model, they demonstrate that providing the *V. cholerae* bacteria in Luria-Bertani (LB) broth results in lethal infection; the flies continually ingest the bacterial suspension and succumb to the pathogen over several days. They further show that toxin co-regulated pili (TCP) do not contribute to virulence, although cholera toxin does play a role [6]. In a later publication, an assay was developed that specifically quantitated colonization [10]. In this approach, *V. cholerae* is fed to flies in phosphate-buffered saline (PBS), a medium that prevents growth of the bacteria, and the bacterial load in the flies' gastrointestinal (GI) tracts is quantified over time by homogenizing the flies and plating serial dilutions onto LB agar supplemented with appropriate antibiotics. To further interrogate colonization dynamics, flies can be transferred from *V. cholerae*-containing broth to sterile media. This washes unattached *V. cholerae* cells from the alimentary canal, and maintenance of the *V. cholerae* population can be quantified over time.

More recently, these techniques have been applied to the development of a laboratory model for colonization of house flies (*Musca domestica*) [13]. A number of historical studies have hinted that house flies could host *V. cholerae* in endemic areas [14–16]. The house fly model provides an opportunity to examine colonization dynamics as *V. cholerae* compete with the flies' diverse commensal microbiota. *V. cholerae* must also contend with house fly immune defense mechanisms, which may have evolved to restrict infection by pathogens ingested from human and animal waste [17].

2 Materials

Safety note. Work with *V. cholerae* should be performed in a laboratory space that complies with relevant biosafety guidelines. A biosafety cabinet may be required, particularly for experiments that involve large volumes of culture or procedures with a high risk of aerosolization. Choosing either O1/O139 strains (engineered to lack the *ctxAB* genes and/or the *tcpA* gene) or non-toxigenic non-O1/non-O139 strains can help mitigate the risk to lab personnel. Thus far, most tested strains of *V. cholerae* have been pathogenic toward *Drosophila*, including those that carry neither the *ctxAB* nor the *tcp* genes, suggesting that the mechanisms of arthropod colonization and infection are conserved.

Investigators should perform experiments under conditions that minimize the risk of infected flies escaping from the assay containers or the lab. The American Committee of Medical Entomology and the American Society of Tropical Medicine and Hygiene have jointly published guidelines for the safe use of insect vectors carrying pathogens [18, 19]. Important safety precautions include separating space for work with *V. cholerae*-infected *Drosophila* from other laboratory space by at least two doorways and a vestibule and covering air vents with appropriately sized mesh. Investigators interested in setting up their own facility for routine infection of *Drosophila* by *V. cholerae* or other pathogens should familiarize themselves with these guidelines and discuss them with their institutional biosafety committees.

2.1 Materials for Drosophila Maintenance

1. Fly vials (Genesee Scientific).
2. Cellulose acetate plugs (Flugs) (Genesee Scientific).
3. Fly bottles (Genesee Scientific).
4. Clear plastic deli containers or insect cups (32 oz) with polyfabric-covered lids (*see* **Note 1**).
5. Fly food: 62.5 g cornmeal, 9.25 g agar, 25.75 g yeast, 62.5 g molasses, 16 mL mold inhibitor (100 g methyl paraben per L 95% ethanol). Add 6 mL acid mix (1.6 L propionic acid, 160 mL phosphoric acid, 2 L H_2O) to each liter of fly food (*see* **Note 2**).

6. Cheesecloth.
7. Dry active baker's yeast (Red Star).
8. Wild-type or mutant fly stock (*see* **Note 3**).
9. Fly "morgue" (500 mL bottle with ~250 mL 70% ethanol or isopropanol).
10. Fly pad (Genesee Scientific).
11. CO_2 tanks, tubing, and regulators for anesthetization, or FlyNap (Carolina Biological Supply) (*see* **Note 4**).
12. CO_2 blowgun (Genesee Scientific).
13. Dissecting microscope (optional).
14. *Drosophila* incubator (*see* **Note 5**).
15. Freezer set at −20 °C.

2.2 Materials for Drosophila Survival Assays

1. Sterile culture tubes.
2. LB agar plates.
3. LB Miller broth.
4. Incubator set at 27 °C or 37 °C with shaker and a shelf.
5. Antibiotics (if necessary).
6. K-resin wide fly vials (Genesee Scientific).
7. Cellulose acetate plugs (Flugs) for wide fly vials (Genesee Scientific).
8. Bandage shears or scissors.
9. Small paintbrush (such as those used for artwork).
10. 15 mL Falcon tubes.

2.3 Materials for Drosophila Colonization Assays

1. Microcentrifuge tubes (1.5 mL), sterile.
2. Benchtop microcentrifuge.
3. Sterile PBS.
4. Vortex.
5. 70% ethanol.
6. Small paintbrush.
7. LB agar plates.
8. Fisherbrand Pellet Pestle Cordless Motor handheld homogenizer (Fisher Scientific).
9. Sterile pestles for homogenizer (Fisher Scientific).
10. 5-Bromo-4-chloro-3-indolyl β-D-galactopyranoside (X-gal, 40 mg/mL in dimethylformamide), for competition colonization assays (*see* **Note 6**).
11. LB agar plates with X-gal, 40 μg/mL, for competition colonization assays.

2.4 Materials for Visualizing Colonization Using a Confocal Microscope

1. Petri dish with Sylgard 184 silicone elastomer (Electron Microscopy Sciences) (*see* **Note 7**).
2. Sterile PBS.
3. Paraformaldehyde (8%).
4. Fine forceps (#5 with biology tips, 0.05 × 0.02 mm, Dumostar; or standard tips, 0.1 × 0.06 mm, Fine Science Tools).
5. Small paintbrush.
6. Nine-well glass plate.
7. Grace's insect medium (optional).
8. 4′,6-Diamidino-2-phenylindole (DAPI; 5 mg/mL in sterile water).
9. VECTASHIELD Mounting Medium.
10. Glass slides.
11. Glass coverslips.
12. Clear nail polish.
13. Dissecting microscope.
14. Confocal microscope.

2.5 Materials for Maintenance of House Flies

1. Colony cages: metal mesh cages or similar secure structures that allow air passage and easy access for changing food and water and measure approximately 0.6 m × 0.6 m × 0.6 m. Adding a cloth sleeve to one side of the cage allows for easy opening and closing while preventing the flies from escaping. It may be possible to maintain small numbers of flies in 32 oz insect cups as for *Drosophila*.
2. Clear plastic deli containers or insect cups (24 oz) with polyfabric-covered lids for house fly diet.
3. Small plastic cups (~5 oz) for water and granulated sugar.
4. Very small plastic cups (~1.5 oz).
5. Granulated sugar.
6. Styrofoam piece (~2.5 cm diameter; to fit in 5 oz plastic cup).
7. Cotton balls.
8. Ammonia.
9. Sand.
10. Plastic tub (~50 cm × ~38 cm × ~13 cm).
11. Mesh, such as insect netting sold by garden supply stores.
12. Large rubber bands.
13. Strainer.
14. House fly artificial diet: 80 g FS wheat bran (PMI Nutritional International LLC), 40 g calf manna nutritional supplement (Manna Pro), and 10 g vermiculite (Vigoro) per 300 mL water in a 24 oz deli container (insect cup).

2.6 Materials for House Fly Colonization Assays

1. Dissecting scissors.
2. 0.25 M sucrose solution, filter-sterilized.
3. Streptomycin solution (50 mg/mL stock in water), filter-sterilized.
4. Kanamycin solution (50 mg/mL stock in water), filter-sterilized.
5. LB agar plates.
6. LB Miller broth.
7. Sterile culture tubes, 125 mL Erlenmeyer flasks, or mini-bioreactors (Corning) (*see* **Note 8**).
8. Oak Ridge centrifuge tubes (Nalgene), sterilized.
9. Sorvall centrifuge.
10. Sterile PBS.
11. Antibiotics, as appropriate.
12. Spectrophotometer.
13. Microcentrifuge tubes (1.5 mL), sterile.
14. Clear plastic deli containers or insect cups (32 oz) with polyfabric-covered lids.
15. Granulated sugar.
16. Surgical cellulose wadding.
17. Tap water, filter-sterilized.
18. Small 1.5 oz cups, plastic.
19. Cellulose acetate plugs (Flugs) for wide fly vials (Genesee Scientific).
20. 15 mL culture tube, plastic.
21. *Drosophila* incubator.
22. 70% ethanol.
23. 10% bleach, freshly prepared.
24. Sterile pure water.
25. Kimwipes.
26. Fisherbrand Pellet Pestle Cordless Motor handheld homogenizer (Fisher Scientific).
27. Sterile pestles for homogenizer (Fisher Scientific).
28. Petri dish with Sylgard 184 silicone elastomer (Electron Microscopy Sciences) (*see* **Note 7**).
29. Sterile PBS.
30. Paraformaldehyde (8%).
31. Fine forceps (#5 with biology tips, 0.05 × 0.02 mm, Dumostar; or standard tips, 0.1 × 0.06 mm, Fine Science Tools).

32. Small paintbrush.
33. Nine-well glass plate.
34. Grace's insect medium (optional).
35. 4′,6-Diamidino-2-phenylindole (DAPI; 5 mg/mL in sterile water).
36. VECTASHIELD Mounting Medium.
37. Glass slides.
38. Glass coverslips.
39. Clear nail polish.
40. Dissecting microscope.
41. Confocal microscope.

3 Methods

3.1 Maintenance of Drosophila melanogaster

1. Place fly food in each insect cup to a depth of ~2 cm (Fig. 1). Cover the food with layered cheesecloth to prevent the accumulation of condensation, and allow the food to dry overnight. Store fly food at 4 °C for up to 2 weeks.
2. Before adding flies to the insect cups, ensure that pools of water, which can trap flies, have not accumulated on the surface of the food. Also, ensure that the food is not too dry, as indicated by separation of the food by the side of the cup. If separation occurs, fresh sterile water can be pipetted directly into the cup to assist larvae in churning the food (*see* **Note 9**).

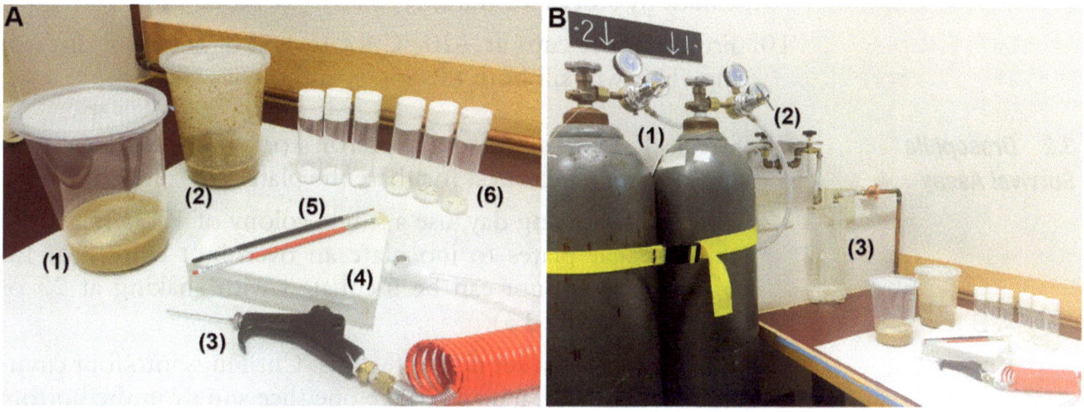

Fig. 1 Equipment for working with *Drosophila* and performing infection assays. (**a**) Flies that have eclosed within the past 4–10 days are placed in 32-oz insect cups with polyfabric lids, with fly food and additional yeast (*1*). Larvae churn the food and climb the side of the cup to pupate (*2*). Flies are anesthetized with CO_2 provided by a blowgun (*3*) and placed on a CO_2 pad (*4*) for manipulation with small paintbrushes (*5*). Flies are added from the CO_2 pad to vials (*6*) for experiments. (**b**) CO_2 is provided by gas cylinders (*1*) and controlled by a CO_2 regulator (*2*). Bubbling CO_2 through water (*3*) can help mitigate static electricity

3. Immediately before adding flies, sprinkle dry active baker's yeast over the top of the fly food in each insect cup to promote female egg-laying (Fig. 1).
4. Transfer the flies to the insect cups at a "fly station," a lab bench that has been equipped with the necessary tools to anesthetize, sort, and maintain *Drosophila* (*see* **Note 10**).
5. To prevent accumulation of bacteria and mold, transfer flies to fresh food regularly. Initiate an insect cup by adding 100–300 adult flies 4–6 days after they have eclosed (hatched from the pupae). Allow the flies to lay eggs, and ensure that the eggs hatch and the larvae successfully churn and soften the food (usually within a week). After the larvae have churned the food and begun migrating up the side of the insect cup to pupate, discard the adult flies into the fly morgue. The new flies will begin eclosing several days later.
6. Transfer the new flies to fresh food within 4 days of eclosing, and allow them to feed on fresh food for an additional 4 days before collecting them for experiments. Flies should be used for experiments 4–10 days after eclosion.
7. Maintain several independent lines of flies, and transfer them to new insect cups on alternating schedules to ensure that flies are available for experiments if one stock becomes sick or contaminated (*see* **Note 11**).
8. When working with flies, wear gloves, wipe down fly equipment with 70% ethanol between work with different fly lines, use sterile tips to protect the nozzle on the CO_2 gun, and maintain a clean incubator and fly room.
9. Check fly stocks for mites regularly (*see* **Note 12**).
10. Freeze insect cups at −20 °C for several days before discarding. This will kill both the adults and the larvae.

3.2 Drosophila Survival Assay

1. Streak *V. cholerae* from a glycerol stock onto LB agar with appropriate antibiotics. Incubate the plates overnight at 37 °C.
2. Late the following day, use a single colony of *V. cholerae* from the LB agar plates to inoculate an overnight culture of LB broth. The culture can be incubated with shaking at 27 or 37 °C (*see* **Note 13**).
3. On the third day, set up the fly vials. Cut Flugs into four circular slices with shears, and lodge one slice snugly in the bottom of each vial to absorb the *V. cholerae* culture. Seal the top of each vial with a whole Flug (Fig. 2). Prepare enough vials so that each *V. cholerae* strain can be tested in triplicate. Include three control vials for uninfected flies, which will contain LB broth without *V. cholerae*.

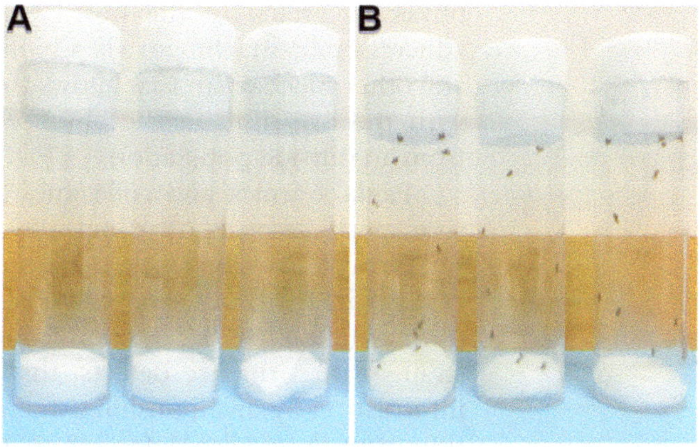

Fig. 2 Vials for performing infection assays with *Drosophila*. (**a**) Vials for assays are prepared by placing a circular section of Flug, cut to a slice of ~1 cm thick, snugly at the bottom of a wide K-resin vial. (**b**) Two milliliters of media with *V. cholerae* is absorbed by the Flug slice, and ten male flies are added to each vial. Survival and colonization are monitored over succeeding days

4. Prepare the required amount of inoculum by diluting overnight *V. cholerae* culture with LB broth (1:10) in a 15 mL Falcon tube (*see* **Note 14**).
5. Mix the inoculum by inverting the Falcon tube several times. Add 2 mL of inoculum to the Flug slice at the bottom of each vial, distributing the culture around the Flug surface. For uninfected control vials, add LB broth without *V. cholerae*. Secure the top of each vial with a Flug. Use secondary containment if the vials are transported to a different room.
6. Perform each assay using flies from a single insect cup (*see* **Note 15**).
7. Empty the insect cup onto a fly pad that dispenses CO_2. Separate the males from the females, and discard the females in the fly morgue (*see* **Note 16**).
8. Add ten male flies to each of the vials (Fig. 2), and place vials in the *Drosophila* incubator at 24 °C with 12-hour light–dark cycles.
9. Monitor fly survival at least twice daily for ~100 h, or until all wild-type flies have died.
10. Graph results using a survival plot, and test for statistical significance using the log-rank test (*see* **Note 17**).
11. Ensure that all flies have died before discarding the vials in an appropriate biohazardous waste container.

3.3 *Drosophila* Colonization Assays

Experimentally, colonization can be defined as the presence of bacteria at a particular location in a host organism. However, a more biologically meaningful definition requires that the bacteria physically attach to the host. This may involve elaboration of adherent pili, attachment via a biofilm, invasion of host tissues, or any other mechanism that allows the bacteria to remain associated with the organism in the absence of continuous replenishment from an outside population.

Paula Watnick and colleagues have developed assays to measure specific aspects of colonization dynamics in fly hosts. In the survival colonization assay (*see* Subheading 3.3.1), flies consume *V. cholerae* in LB broth, and bacterial loads are measured over time during the course of lethal infection. When used with *V. cholerae* strains with virulence defects, this assay can provide mechanistic information by distinguishing strains with deficiencies in colonization from those with defects in toxicity or other damage-inducing pathways. However, interpretation of the results can be complicated by the fact that flies ingest *V. cholerae* throughout the assay, continuously replenishing the pathogen population in their GI tracts. To examine colonization in the absence of continued ingestion, Purdy and Watnick (2011) developed the colonization transfer assay (*see* Subheading 3.3.2). In this assay, *V. cholerae* are fed to flies in PBS medium, which does not support growth of *V. cholerae* and renders them nonlethal. After feeding on *V. cholerae* in PBS medium for 48 h, the flies are transferred to fresh PBS medium without *V. cholerae*. Colonization can be measured by homogenizing whole flies and counting bacterial loads or by using confocal microscopy to observe the distribution of a green fluorescent protein (GFP)-expressing strain of *V. cholerae* in the dissected alimentary canal. A competition colonization transfer assay (*see* Subheading 3.3.3) enables further exploration of the colonization process. Two strains with different *lacZ* phenotypes are fed to the flies simultaneously, and their ability to compete during colonization is measured using counts of blue and white colonies on plates with the colorimetric substrate X-gal.

3.3.1 Survival Colonization Assay

1. Set up assays as described for *Drosophila* survival (*see* Subheading 3.2), but prepare three vials per *V. cholerae* strain for each time point at which you wish to measure bacterial loads.
2. At designated time points (usually 24 h, 48 h, and sometimes 72 h after infection), anesthetize flies with CO_2, and add them directly to 1.5 mL microcentrifuge tubes containing 300 μL of sterile PBS. Accomplish this as follows:
   (a) Cut the end (~0.25 cm) off of a 200 μL pipette tip, and place the remaining tip on the end of the CO_2 blowgun.
   (b) Loosen the top Flug of the fly vial slightly, and gently insert the pipette tip between the Flug and the inside rim of the vial. If all flies are alive, tip the vial downward, so that it is almost fully inverted. The *V. cholerae* culture, which has saturated the Flug, will not drip down the vial.

If one or more flies in the vial have died, tip the vial slightly, and ensure that the dead flies remain stuck in the bottom of the tube. Gently add CO_2 to the vial through the pipette tip, and allow the live flies to fall onto the inside of the top Flug. When the flies are fully anesthetized, remove the CO_2 blowgun, and discard the pipette tip.

(c) Hold the vial horizontally, and remove the top Flug, being careful to ensure that all flies remain in the vial. Using a small paintbrush, gently sweep the flies into a labeled 1.5 mL microcentrifuge tube with 300 μL of sterile PBS. Only transfer live flies, as dead flies are likely to carry high bacterial loads. Record the number of flies added to each tube. Keep the microcentrifuge tubes on ice to immobilize the flies.

(d) Place a new tip on the CO_2 blowgun between each vial transfer. Rinse the paintbrush with 70% ethanol between each vial transfer and at the end of the procedure.

3. Homogenize the flies in PBS using a handheld homogenizer and sterile pestles until their abdomens and thoraxes are completely disrupted (*see* **Note 18**). Though the tough exoskeleton may not be completely disintegrated, the GI tract should be fully homogenized. Place the tube on ice and allow the large fly parts to settle for 2 min.
4. Transfer 50 μL of the supernatant to 450 μL of sterile PBS, and perform serial dilutions to 10^{-5} (*see* **Note 19**).
5. Spread 100 μL of each dilution onto appropriate agar plates (use only 50 μL from the tube in which flies were homogenized, since little volume will be available after the fly parts settle). If the *V. cholerae* strains carry an antibiotic resistance marker, then spreading the dilutions onto LB plates with antibiotics will suppress growth of contaminating bacteria. If not, the endogenous bacteria of the fly are typically more slow-growing than *V. cholerae* and are easily distinguished by colony morphology (*see* **Note 20**).
6. Count colonies on plates with 30–300 colonies. Bacterial load is calculated as (# colonies × dilution factor × 300 μL)/(volume on plate in μL × number of flies).

3.3.2 Colonization Transfer Assay

1. Streak *V. cholerae* strains, and set up overnight cultures as described above (*see* Subheading 3.2, **steps 3** and **4**). A total of 2.4 mL of overnight culture will be needed.
2. After overnight cultures have grown, set up fly vials for experiments (*see* Subheading 3.3.1, **step 1**). Test the bacterial load of each strain at 24 and 48 h and again after 72 h without transfer and 72 h with transfer. This requires four sets of three vials per each bacterial strain.

3. Place 1.2 mL of overnight culture in each of two microcentrifuge tubes and spin at 2300 × *g* for 6 min. Discard the supernatant and resuspend the bacterial pellet in an equal volume of sterile PBS. The PBS can be supplemented with antibiotics if the *V. cholerae* carry antibiotic resistance markers.
4. Dilute the overnight cultures 1:10 in PBS to obtain a total of 2 mL of diluted culture per vial.
5. Add 2 mL of diluted culture to the Flug at the bottom of each vial, spreading the culture around the Flug as it is being added. Cap the vial with another Flug.
6. After the *V. cholerae* has been added to each of the vials, add ten male flies, and place the vials in the *Drosophila* incubator.
7. After 24 and 48 h, homogenize flies to measure bacterial loads, as described above (*see* Subheading 3.3.1, **steps 2–6**).
8. After 48 h, add sterile PBS to sets of three vials for each bacterial strain being tested. To transfer the flies, carefully anesthetize them with CO_2, and use a small paintbrush to push them into the vial with sterile PBS. Be careful to only transfer live flies at this step.
9. After 72 h on *V. cholerae*, or 48 h on *V. cholerae* and 24 h off of *V. cholerae*, flies can be tested for bacterial load as described above (*see* Subheading 3.3.1, **steps 2–6**).

3.3.3 Competition Colonization Transfer Assay

1. Experiments are set up as described above (*see* Subheading 3.3.2, **steps 1–9**), except that *V. cholerae* strains with *lacZ*$^+$ and *lacZ*$^-$ alleles are used.
2. Prepare X-gal plates by adding X-gal solution to autoclaved LB agar to a final concentration 40 μg/mL. Upon homogenization and plating, blue and white colonies can be easily distinguished on X-gal plates.
3. If testing a wild-type *V. cholerae* strain against a mutant strain, then complementary experiments should be performed in which both the WT/*lacZ*$^+$ and mutant/*lacZ*$^-$ strains, as well as the WT/*lacZ*$^-$ and mutant/*lacZ*$^+$ strains, are compared to each other, to be sure that *lacZ* has no effect on the phenotype.

3.3.4 Visualizing Colonization in the Confocal Microscope

1. Perform experiments as described previously (*see* Subheading 3.3.1). Use *V. cholerae* carrying a plasmid- or chromosome-borne marker that expresses a fluorescent protein such as GFP.
2. Anesthetize flies with CO_2 and place in a microcentrifuge tube on ice (*see* Subheading 3.3.1).
3. Place 50–100 μL of PBS in a petri dish containing silicone elastomer. Using forceps, submerge one to two flies in the PBS droplet.

4. Under a dissecting microscope, against a black background, use fine forceps to immediately remove the head, wings, and legs of a fly.
5. Gently pull apart the abdomen and thorax, and separate the alimentary canal from other internal organs, including the testes and Malpighian tubules. Retain the crop and hindgut if possible, so that the entire length of the alimentary canal can be examined for colonization.
6. Add 300 μL of PBS or Grace's medium to each well of a nine-well glass plate, and place the entire plate on ice.
7. Transfer the GI tract to the nine-well glass plate, keeping the plate on ice.
8. As soon as possible, transfer the GI tracts to 250 μL PBS/250 μL 8% paraformaldehyde under a fume hood. Fix the GI tracts for 30 min.
9. Wash three times briefly by carefully transferring each GI tract to a new well with PBS.
10. Wash twice for 30 min. in 500 μL PBS at 4 °C on ice or in a cold room.
11. Use 1 μg/mL DAPI in the last wash and incubate in the dark (*see* **Note 21**).
12. To mount the GI tracts, place 2 μL of VECTASHIELD on a glass slide. Carefully transfer the GI tract, and untangle the tract as much as possible without damaging it. Place a coverslip over the GI tract, taking care to avoid air bubbles. Seal coverslips with clear nail polish. Store slides at 4 °C until ready to visualize on the confocal microscope.

3.4 Maintenance of Musca domestica

House flies are an important vector of human pathogens, particularly because they reside together with humans, often in areas where waste accumulates [20]. We have recently developed a model of *V. cholerae* house fly infection based on protocols designed for gauging colonization in *Drosophila* [13]. Our initial findings indicate that *V. cholerae* colonization dynamics in the two fly species differ substantially, providing an opportunity to examine a new set of parameters that modulate *V. cholerae* interactions with arthropod species. Unlike its colonization of *Drosophila*, *V. cholerae* does not elaborate an adhesive biofilm-like structure in the house fly hindgut, and *V. cholerae* appears to encounter a greater bottleneck to infection in the house fly, with extensive variation in colonization levels among individual flies. This model provides a novel opportunity to explore dynamics that control colonization in an ecologically relevant host and will allow for further evaluation of factors that control vectoring of *Vibrio* by arthropods.

It is possible to maintain a house fly colony yourself, or you may order house fly larvae or pupae from Carolina Biological Supply, as well as house fly medium and a house fly culturing kit. Instructions for maintaining house flies are provided below, based on protocols generously provided by John Stoffolano, University of Massachusetts Amherst.

1. Maintain house flies in mesh cages or 32 oz insect cups. Provide a continuous light cycle of 16 h light to 8 h dark, a temperature of 27 °C, and 50% relative humidity. Provide the flies with tap water and a Styrofoam floater to prevent drowning. Also provide small cups of granulated sugar.
2. Prepare house fly artificial diet by first crushing the calf manna, and then mixing in the wheat bran and the vermiculite. Add water until the mixture is moistened (~300 mL). Place the diet into a 24 oz plastic deli container.
3. To collect house fly eggs, first moisten a cotton ball with ammonia, and place the cotton ball into a 32 oz plastic insect cup. Cut a 2-cm diameter hole in the lid of the insect cup. Place a small plastic cup (1.5–2 oz) in the hole so that it perches in the lid. Place wheat bran in the small plastic cup, and moisten the wheat bran with tap water. Place the entire cup (the large cup with cotton ball, together with the small cup in the lid with the wheat bran) into the colony cage overnight. The following morning, collect the eggs from the wheat bran using a metal spatula, and place them into a 24 oz plastic deli container with house fly artificial diet. Bury the eggs just under the surface in the damp diet.
4. To collect house fly pupae, place the cup with house fly artificial diet and eggs into a plastic tub with sand in the bottom. Cover the top of the tub with mesh and rubber bands. After a week, the house fly larvae will crawl out of the diet and into the sand, where they will pupate.
5. Collect the pupae from the sand using a strainer. Place the pupae in an open cup, and put them into the colony cage, where they will emerge.

3.5 House Fly Colonization Assays

3.5.1 Preparation of M. domestica and V. cholerae for Colonization Assays

1. As flies emerge, separate the males, and truncate their wings with scissors (*see* **Note 22**). Transfer the males to a new cage. The females may be discarded or kept for further breeding.
2. Provide male flies with filter-sterilized 0.25 M sucrose for ~18 h. Remove the sucrose, and provide the flies autoclaved or filter-sterilized water overnight.
3. The following day, provide flies with 0.25 M sucrose supplemented with high concentrations of antibiotics (streptomycin at 5 mg/mL and kanamycin as 10 mg/mL) [21]. Allow the flies to feed on the sucrose/antibiotic solution for 18 h.

Again remove the solution and provide the flies with water but no sucrose, food, or antibiotics overnight.

4. On the third day, maintain flies on water and begin the experiments.
5. To prepare *V. cholerae*, streak *V. cholerae* strains onto LB agar plates (with appropriate antibiotics if the strain carries a plasmid that confers GFP expression) 2 days before the start of the experiment.
6. The following day, inoculate 24 mL of LB broth in an appropriate culture vessel (*see* **Note 8**), and incubate overnight with shaking at 200 rpm.
7. Pellet the culture in sterile Nalgene Oak Ridge tubes by spinning in a Sorvall centrifuge using an SS-34 rotor at 4500 × *g* for 5 min at 4 °C. Discard supernatants, and resuspend cells in 8 mL of sterile PBS.
8. If the *V. cholerae* strain is carrying a plasmid for GFP expression, add the appropriate antibiotics to help ensure that the plasmid is maintained throughout the experiment.
9. Measure the OD_{600} of the concentrated bacterial suspension (*see* **Note 23**).
10. To measure bacterial numbers in the culture, perform serial dilutions of the concentrated bacterial suspension (*see* Subheading 3.3.1, **steps 4–6**).
11. Repeat this procedure on each day of the experiment, in order to provide the flies with fresh bacteria daily.

3.5.2 Performing House Fly Colonization Assays

1. Set up 32 oz plastic deli (insect) cups, used for maintaining *Drosophila*, with a small cup of granulated sugar, a cup with surgical cellulose wadding saturated with filter-sterilized water, and a cup with a Flug cut in half to a dimension of 2.6 cm diameter × 1.3 cm depth (*see* **Note 24**).
2. Saturate the Flug with ~4 mL of bacterial preparation.
3. Add 4–6 male flies to each insect cup. Catch the flies in their cages using a 15 mL culture tube, and quickly deposit them in the insect cup. Anesthetization may not be required.
4. Place the cups in the *Drosophila* incubator.
5. Perform quantitation of bacterial loads as described above for *Drosophila* (*see* Subheading 3.3.1, **steps 2–6**), except for the following changes:
   (a) Surface sterilize the flies by placing them in a 15 mL tube with 10 mL of 70% ethanol for 10 min immediately after collecting them. Then rinse the flies in sterile water, and transfer them to a second 15 mL tube with 10 mL of freshly prepared 10% bleach for 10 min. Then rinse them once

more in sterile water. Gently blot each fly with a fresh Kimwipe to remove excess water, and place them individually in microcentrifuge tubes with 100 μL of sterile PBS on ice.

(b) Homogenize house flies individually. It may be prudent to work in a biosafety cabinet due to the risk of aerosolization.

(c) Prior to performing serial dilutions, add 200 μL of PBS so that the supernatant can be pipetted away from the settling fly parts.

6. Perform dissection and confocal microscopy as described above for *Drosophila* (*see* Subheading 3.3.4).

4 Notes

1. While small numbers of *Drosophila* are typically raised in fly vials or bottles (available from Genesee Scientific), it may be more convenient to raise large numbers of flies in 32 oz clear plastic deli (insect) cups with polyfabric-covered lids. These cups are available from stores that supply products for maintenance of insects, amphibians, and reptiles.
2. There are many different fly food recipes (*see* http://flystocks.bio.indiana.edu/Fly_Work/media-recipes/media-recipes.htm). We have found this cornmeal-molasses-yeast recipe to give consistent results. For labs without access to a fly food facility, it may be convenient to purchase premixed fly food from Genesee Scientific.
3. Stocks of flies can be requested from the Bloomington *Drosophila* Stock Center at Indiana University and other stock centers around the world. Common wild-type stocks include Oregon-R and Canton-S. If mutant flies are used, the appropriate wild type for that strain should be examined simultaneously.
4. Flies are most easily and safely manipulated when they are anesthetized with CO_2. The CO_2 gas is typically stored in a cylinder with an appropriate regulator and piped directly to stations with CO_2 pads (for sorting flies) and a CO_2 blowgun (for anesthetizing flies in containers). Passage of CO_2 through a bubbler (a sealed container that CO_2 enters through a short tube submerged in water and leaves through piping at the top) helps mitigate issues with static electricity. In the absence of a system for providing CO_2, it may be possible to use FlyNap (Carolina Biological Supply) or short exposures to cold temperature to anesthetize flies.
5. Flies are raised in an incubator with regular light–dark cycles of 12 h at a temperature of 24 °C. Although humidity is not

necessary, it may be important in buildings or climates with particularly dry air. Adding bins of water to the bottom of the incubator has been sufficient to maintain adequate levels of humidity, but more drastic measures may be needed in drier environments. Percival (Perry, IA) and Genesee Scientific (San Diego, CA) both supply incubators specifically for raising *Drosophila*.

6. Do not use polystyrene pipettes to pipette dimethylformamide. Use glass or polyethylene pipettes instead. Store X-gal at −20 °C protected from light (wrapped in aluminum foil).
7. To make a petri dish for dissection, mix Sylgard silicone according the manufacturer's instructions, dispense into the bottom of a plastic petri dish until the plate is ~3/4 full, and allow it to solidify.
8. To culture a large volume of *V. cholerae* with aeration, we have used disposable conical tubes with 0.2 μm filters in their lids. These are made by Corning and sold as "mini-bioreactors." Alternatively, 50 mL conical tubes with lids fitted loosely, but securely, are an option. Small 125 mL Erlenmeyer flasks would also make appropriate culturing vessels.
9. One of the most important considerations when raising *Drosophila* is to ensure that the food is sufficiently wet to support larval growth and maturation. Food moisture can be adjusted as the food is being made, and water can also be added after the food is set. Insect cups should be monitored regularly to ensure that the food does not separate from the side of the insect cup, an early indicator of dryness.
10. A permanent fly station is usually set up at a table-height benchtop and outfitted with CO_2 pads and a CO_2 blowgun. A dissecting microscope is necessary when screening for fly genetic markers, but may not be necessary when working with wild-type flies. Use of a dissecting scope can be helpful when learning to sex wild-type flies for the first time.
11. On occasion, fly populations can exhibit a tendency to die on LB medium in the absence of *V. cholerae*. We hypothesize that flies can become colonized with opportunistic pathogens that, when inoculated into LB broth, grow to high concentrations and lethally infect the flies (P. I. Watnick, personal communication). In the past, this has occurred after an extended period of low humidity when insect cups had been allowed to desiccate. Efforts to treat the flies with antibiotics and daily transfers onto fresh food did not solve the problem, and stocks were ultimately discarded. The addition of streptomycin (100 μg/mL) to LB broth, when using streptomycin-resistant *V. cholerae* strains, may also mitigate this issue.

12. Mites can be a nuisance in fly labs and should be avoided when studying infection and immunity in *Drosophila*. Practice good mite control by quarantining all new fly stocks in a region of the lab away from the lab's current stocks, and check for mites under the dissecting microscope before introducing the new stock into the lab's incubator. Mites are only visible under a dissecting microscope, and they can be more apparent after fly vials are several weeks old.
13. In our laboratory, we have set up overnight cultures at both 27 and 37 °C, without evident effect on virulence toward *Drosophila*. However, one temperature should be maintained for all assays.
14. In our lab, normalizing cultures by optical density has not been necessary at this step. However, it may be prudent to normalize cultures of strains with different growth rates. In addition, it may be particularly important to measure colonization in the fly, to see if growth defects contribute to possible virulence defects. Antibiotics such as streptomycin can be added at this step.
15. On occasion, different batches of flies have shown variable levels of susceptibility to *V. cholerae*. To minimize variability, flies originating from a single insect cup should be used for each experiment. If one insect cup does not yield sufficient flies, then two or three cups may be used, but appropriate controls should be repeated in each fly cup. For example, negative controls (flies fed LB without *V. cholerae*) and positive controls (flies fed wild-type *V. cholerae*, or another appropriate positive control strain for the assay) should be tested with flies from each insect cup.
16. Males are smaller than females and have darker coloration at the posterior region of their abdomen. Only male flies are used for experiments, since females have a more variable response to *V. cholerae* infection that often results in slower death (P. I. Watnick, personal communication). Throughout, CO_2 should be maintained at the lowest level that keeps the flies asleep, since extended exposure to high levels of CO_2 can weaken the flies and cause them to die soon after the start of the experiment. If several hundred flies are being sorted at once, it may be beneficial to sort the males and females, discard the females, and then transfer the males to fresh fly food and allow them to awaken. An hour or so later, the flies can be anesthetized again before dividing them into groups of ten and placing them into fly vials.
17. GraphPad Prism is a convenient tool for simple statistical analysis.

18. Given the risk of aerosolization at this step, it may be prudent to work in a biosafety cabinet.
19. Keep the homogenized flies on ice, and perform the initial 1:10 dilution immediately to prevent growth of the *V. cholerae* in the fly homogenate.
20. Controls should be performed to ensure that stochastic colonization with a strain such as *Pseudomonas*, which is difficult to distinguish from *V. cholerae* after overnight growth, has not occurred. These include homogenizing and plating vials of uninfected flies. To further ensure that colonies are in fact *V. cholerae,* patch individual colonies onto thiocitrate bile salts (TCBS) agar and incubate overnight at 37 °C. Colonies that appear yellow on TCBS media are most likely *V. cholerae.*
21. DAPI is not soluble in PBS. To add DAPI to the PBS wash, make a 1:100 dilution of the 5 mg/mL stock solution in sterile, pure water, and then use 20 μL of this dilution per 1 mL of PBS.
22. The flies' wings can be removed to help prevent escape and to hinder the movement of an infected fly should it escape. Transferring flies at the start of the experiment is much easier when they lack wings and does not require anesthetization. To minimize the impact on the flies, truncate the wings close to the base without damaging cuticle. Some researchers prefer to keep the flies' wings intact out of concern that wing removal may alter the progression of the flies' interaction with the pathogen by causing stress or altering the strength or severity of the immune response.
23. In order to take an accurate reading, perform a 1:10 dilution of the suspension in PBS before measuring absorbance.
24. The previous experiments did not provide protein in the fly diet. It may be prudent to experiment with different diets to see whether nutrients affect regurgitation or movement of the meal into the midgut.

Acknowledgments

We would like to thank Scott Keating and John Stoffolano (University of Massachusetts Amherst) for generously providing their protocols for maintaining house fly colonies and Ethan Graf for commenting on the manuscript and sharing the fly food protocol. Many thanks to Paula Watnick for her insights and discussions throughout a postdoctoral fellowship in her laboratory that introduced me to the *Drosophila* model of *Vibrio cholerae* infection, and for her continued support.

References

1. Bonfini A, Liu X, Buchon N (2016) From pathogens to microbiota: how *Drosophila* intestinal stem cells react to gut microbes. Dev Comp Immunol 64:22–38. https://doi.org/10.1016/j.dci.2016.02.008
2. Dobson AJ, Chaston JM, Newell PD et al (2015) Host genetic determinants of microbiota-dependent nutrition revealed by genome-wide analysis of *Drosophila melanogaster*. Nat Commun 6:6312. https://doi.org/10.1038/ncomms7312
3. Broderick NA, Buchon N, Lemaitre B (2014) Microbiota-induced changes in *Drosophila melanogaster* Host gene expression and gut morphology. MBio 5:e01117. https://doi.org/10.1128/mBio.01117-14
4. Graczyk TK, Knight R, Gilman RH, Cranfield MR (2001) The role of non-biting flies in the epidemiology of human infectious diseases. Microbes Infect 3:231–235
5. Greenberg B (1973) Flies and disease. Princeton University Press, Princeton, NJ
6. Blow NS, Salomon RN, Garrity K et al (2005) *Vibrio cholerae* infection of *Drosophila melanogaster* mimics the human disease cholera. PLoS Pathog 1:e8. https://doi.org/10.1371/journal.ppat.0010008
7. Berkey CD, Blow N, Watnick PI (2009) Genetic analysis of *Drosophila melanogaster* susceptibility to intestinal *Vibrio cholerae* infection. Cell Microbiol 11:461–474. https://doi.org/10.1111/j.1462-5822.2008.01267.x
8. Guichard A, Cruz-Moreno B, Aguilar B et al (2013) Cholera toxin disrupts barrier function by inhibiting exocyst-mediated trafficking of host proteins to intestinal cell junctions. Cell Host Microbe 14:294–305. https://doi.org/10.1016/j.chom.2013.08.001
9. Wang Z, Hang S, Purdy AE, Watnick PI (2013) Mutations in the IMD pathway and mustard counter *Vibrio cholerae* suppression of intestinal stem cell division in *Drosophila*. MBio 4:e00337–e00313. https://doi.org/10.1128/mBio.00337-13
10. Purdy AE, Watnick PI (2011) Spatially selective colonization of the arthropod intestine through activation of *Vibrio cholerae* biofilm formation. Proc Natl Acad Sci 108:19,737–19,742. https://doi.org/10.1073/pnas.1111530108
11. Hang S, Purdy AE, Robins WP et al (2014) The acetate switch of an intestinal pathogen disrupts host insulin signaling and lipid metabolism. Cell Host Microbe 16:592–604. https://doi.org/10.1016/j.chom.2014.10.006
12. Soler-Bistué A, Mondotte JA, Bland MJ et al (2015) Genomic location of the major ribosomal protein gene locus determines *Vibrio cholerae* global growth and infectivity. PLoS Genet 11:e1005156. https://doi.org/10.1371/journal.pgen.1005156
13. El Bassiony GML, Luizzi V, Nguyen D et al (2016) Laboratory infection of the adult house fly (*Musca domestica*) by *Vibrio cholerae*. Med Vet Entomol 30:392–402
14. Fotedar R (2001) Vector potential of houseflies (*Musca domestica*) in the transmission of *Vibrio cholerae* in India. Acta Trop 78:31–34
15. Gill CA, Lal RB (1931) The epidemiology of cholera, with special reference to transmission. A preliminary report. Indian J Med Res 18:1255–1297
16. Oo KN, Sebastian AA, AYE T (1989) Carriage of enteric bacterial pathogens by house flies in Yangon, Myanmar. J Diarrhoeal Dis Res 7:81–84
17. Scott JG, Warren WC, Beukeboom LW et al (2014) Genome of the house fly, *Musca domestica* L., a global vector of diseases with adaptations to a septic environment. Genome Biol 15:466. https://doi.org/10.1186/s13059-014-0466-3
18. (2003) Arthropod containment levels (ACLs). Vector-Borne Zoonotic Dis 3:75–90. https://doi.org/10.1089/153036603322163475
19. (2003) Risk assessment for arthropod vectors. Vector-Borne Zoonotic Dis 3:69–73. https://doi.org/10.1089/153036603322163466
20. Greenberg B (1971) Flies and disease. Princeton University Press, Princeton, NJ
21. Bansal R, Hulbert S, Schemerhorn B, Reese JC, Whitworth RJ, Stuart JJ, Chen M-S, Ho PL (2011) Hessian fly-associated bacteria: transmission, essentiality, and composition. PLoS ONE 6(8):e23170

Chapter 9

Danio rerio as a Native Host Model for Understanding Pathophysiology of *Vibrio cholerae*

Kristie C. Mitchell and Jeffrey H. Withey

Abstract

Vibrio cholerae is an aquatic bacterium that causes the human disease cholera as well as milder forms of diarrhea. *V. cholerae* is found in the environment in association with a variety of aquatic animals, including vertebrate fish. Here we describe the use of zebrafish (*Danio rerio*) as a model for the pathogenic life cycle of *V. cholerae*. Being that fish are natural hosts for *V. cholerae*, the model provides several significant advantages over existing mammalian models that are not natural hosts.

Key words Cholera, *Vibrio cholerae*, Diarrhea, Zebrafish, Infectious disease transmission

1 Introduction

The zebrafish, *Danio rerio*, has been used extensively in biological research over the past four decades. Due to the transparency of embryos, zebrafish development was initially a fruitful area of study. Zebrafish have also been used in studies of nearly every other aspect of biology, including neuroscience, behavior, cancer, immunology, and infectious diseases, among others [1–3].

Vibrio cholerae is a particularly pertinent human pathogen that has recently been studied using zebrafish as a model host [4]. *V. cholerae* is an aquatic bacterium that is primarily known for causing the deadly human diarrheal disease cholera [5]. Cholera has been endemic in southern Asia throughout recorded history. The native habitat of zebrafish is also southern Asia, so *V. cholerae* and *D. rerio* have likely interacted in their aquatic habitats for thousands of years. Recent work has described several species of vertebrate fish as apparently natural *V. cholerae* hosts [6–9]. However, it is unknown whether *V. cholerae* causes any pathology in these fish. Zebrafish were explored as a potential natural host model and found to recapitulate the *V. cholerae* pathogenic life cycle: (1) fish are infected simply by exposure in water; (2) fish are specifically

Aleksandra E. Sikora (ed.), *Vibrio Cholerae: Methods and Protocols*, Methods in Molecular Biology, vol. 1839,
https://doi.org/10.1007/978-1-4939-8685-9_9,

colonized in the intestine; (3) colonization leads to diarrhea; (4) and *V. cholerae* excreted by infected fish transmit the infection to naïve fish [4].

Here we describe the details behind this model system, with specific information about colonization, dissection, bacterial counts, and transmission experiments.

2 Materials

All dissection tools were obtained from Fine Science Tools unless otherwise specified.

1. Two Dumont #5 forceps.
2. Student Vannas spring scissors.
3. # 3 scalpel handle.
4. Insect pins, size 1.
5. # 11 scalpel blades, sterile.
6. Kelly hemostats.
7. Cling wrap for hemostats.
8. Plastic spoons.
9. Dissecting mat or flat piece of Styrofoam.
10. 2 mL tubes with O-ring caps.
11. 1 mm glass beads.
12. Mini-Beadbeater-24 (BioSpec Products, Inc.).

2.1 Infection/Tank Water

Tap water is passed through reverse osmosis (RO) and then conditioned with 60 mg/L Instant Ocean salt. Infection water is autoclaved before use in experiments.

2.2 Tricaine

Tricaine (ethyl 3-aminobenzoate methanesulfonate salt; Sigma A5040) stock solution is 4 mg/mL in RO water, pH adjusted to 7.0 with 1 M Tris–HCl, pH 9.0.

2.3 Zebrafish

Wild-type, outbred zebrafish, aged 6 months to 1 year, are used for intestinal colonization and bacterial transmission experiments. For anesthesia, zebrafish are placed in 100 mL of 168 μg/mL tricaine solution for approximately 1–2 min or until gill movement slows and fish stop swimming. Fish are recovered from anesthesia by removing them from the Tricaine solution and immediately placing them in fresh infection water. For euthanasia of zebrafish, the dose of tricaine is doubled, and fish remain in the solution for at least 25 min. All animal protocols were approved by the Wayne State University IACUC committee.

3 Methods

3.1 Inoculation of Zebrafish via Immersion

Bacterial cultures are grown with aeration in LB broth at 37 °C for 16–18 h. Cells are subsequently washed once with sterile 1× PBS and then diluted to the correct concentration in sterile 1× PBS. Bacterial cell densities should range from 10^7 to 10^{10} per beaker depending on the desired multiplicity of infection (~5 × 10^4–5 × 10^7 cfu/mL). Four to five zebrafish are placed into a 400 mL beaker containing 200 mL of sterile infection water (Fig. 1). 1000 μL of bacterial inoculum is then added to the beaker with fish, and the beaker is covered with a perforated lid. Each beaker is labeled and placed into a glass-front incubator set at 28 °C for the duration of the experiment.

3.2 Dissection

Scoop a fish out of the tricaine solution with a plastic spoon and place on the dissecting surface. Position the fish with its ventral side facing upward, and pin it through the lower jaw, angled away from the body lengthwise (Fig. 2a). Place another pin just posterior to the anus, also angled away from the body lengthwise (Fig. 2b). Swab the ventral surface with a Kimwipe soaked in 70% ethanol and flame sterilize the scalpel and Vannas scissors. Use the scalpel

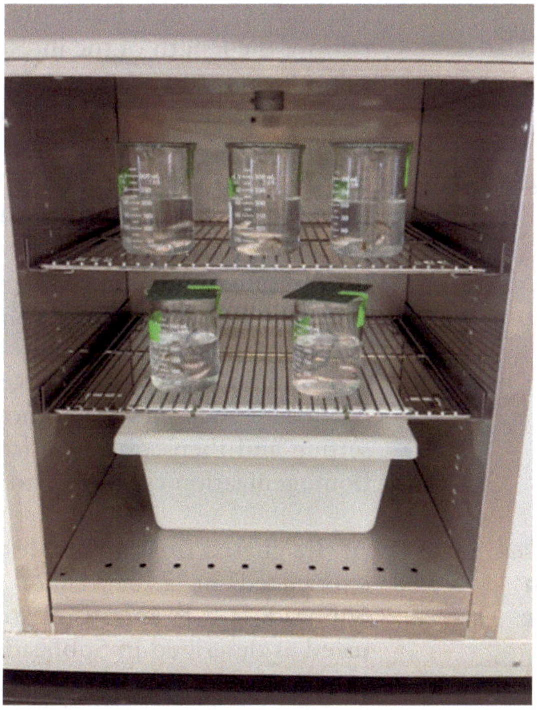

Fig. 1 Incubation of infected zebrafish

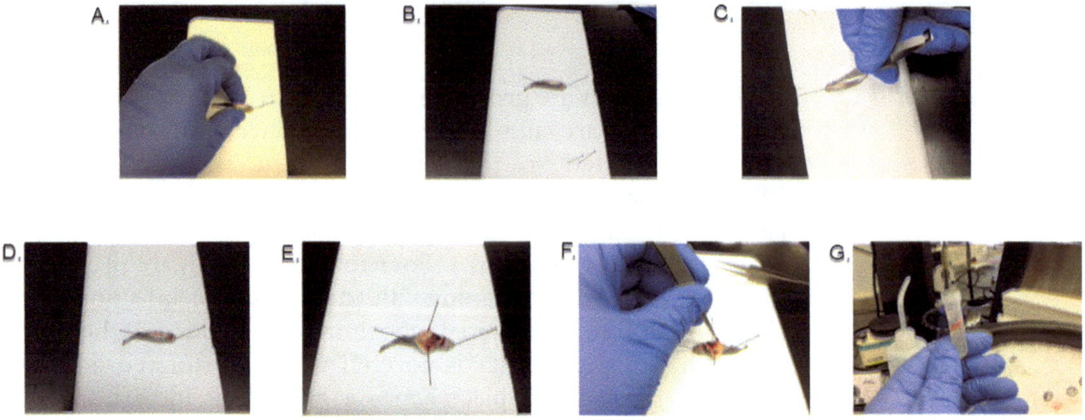

Fig. 2 Dissection of zebrafish intestine (**a**) Pin fish through lower jaw. (**b**) Place second pin posterior to anus. (**c**) Use scalpel to make small incision just posterior to gills. (**d**) Use scissors to extend incision to anal area. (**e**) Pin skin on either side of fish to expose internal organs. (**f**) Use forceps to remove entire length of intetsinal tract. (**g**) Place intestine in screw cap homogenization tube for further processing

to make a small incision in the belly, only penetrating just under the skin (Fig. 2c). Using the scissors, extend the incision along the length of the body, being careful to cut no deeper than skin level and avoiding the anus (Fig. 2d). Pin the skin on each side of the incision to the dissecting surface, angling the pins out away from the body (Fig. 2e). Flame sterilize the forceps, then use them to remove the length of the intestinal tract (Fig. 2f), and place it into a homogenization tube, containing glass beads (*see* below) and 1000 μL of 1× PBS (Fig. 2g), on ice.

3.3 Homogenization

Homogenization tubes are prepared as follows: 2 mL tubes are filled with 1.5 g of 1 mm glass beads (approximately half-way), O-ring caps are loosely screwed on, and then tubes are autoclaved for 20 min. Once cooled, caps should be screwed on tightly and tubes can be stored until needed.

Once zebrafish intestines have been added to the tubes, the caps should be screwed on very tightly and then secured in the beadbeater. Samples are homogenized for 1 min on the maximum setting and then cooled on ice for 1–2 min, before repeating the homogenization cycle once more.

3.4 Intestinal Colonization Experiments

At designated time points, fish are removed from the beaker and euthanized in a beaker of tricaine solution as described in Subheading 2.3. Intestines are aseptically removed and homogenized as described in Subheadings 3.2–3.3. Serial dilutions of the homogenate in LB are made (typically tenfold dilutions) and plated on LB agar plates containing 100 μg/mL streptomycin and 40 μg/mL X-gal. Plates are incubated at 30 °C for 16–18 h, after which *V. cholerae* colonies are counted (*see* **Note 1**). *V. cholerae* typically produces pale blue colonies.

3.5 Transmission Experiments

A group of four to five zebrafish, anesthetized and fin-clipped for later identification (*see* **Note 2**), are placed in a 600 mL beaker with 300 mL sterile infection water. A second group of four to five zebrafish are exposed to 10^7–10^9 *V. cholerae* in 200 mL sterile infection water as described above for 2–3 h to establish colonization. If a control for colonization at this time point is desirable, one or two fish can be sacrificed and dissected and intestinal homogenates plated as described above. The remaining infected fish are then transferred to another beaker of fresh infection water two times sequentially to remove external *V. cholerae*. The rinsed, infected fish are placed in the beaker already holding the fin-clipped naïve zebrafish. After ~24 h all the fish are sacrificed, and intestinal *V. cholerae* are enumerated as described above, separating the fin-clipped and unclipped fish.

4 Notes

1. If multiple colony morphologies are observed, this is likely due to natural streptomycin resistant microbiota. To verify bona fide *V. cholerae*, patch onto TCBS plates, on which *V. cholerae* produces large yellow colonies. Usually with a bit of experience, the *V. cholerae* colony morphology is easy to distinguish; however, different *V. cholerae* strains have different colony morphologies, so it is best to test each morphological variant on TCBS when first performing an experiment. Non-O1/O139 *V. cholerae* can be particularly tricky to identify without significant experience.
2. For fin clipping, fish are anesthetized as described in Subheading 2.3, and the tail fin is clipped with scissors that have been flamed to sterilize. Just enough tissue is removed to make identification clear from unclipped fish. Clipped fish are then returned to the beaker of water to recover.

Acknowledgments

This work was supported by NIH grants R21AI095520 and R01AI127390 and funding from Wayne State University. We also thank Dr. Melody Neely and her laboratory, who helped in developing this model.

References

1. Allen JP, Neely MN (2010) Trolling for the ideal model host: zebrafish take the bait. Future Microbiol 5(4):563–569
2. Rowe HM, Withey JH, Neely MN (2014) Zebrafish as a model for zoonotic aquatic pathogens. Dev Comp Immunol 46(1): 96–107
3. Sullivan C, Kim CH (2008) Zebrafish as a model for infectious disease and immune function. Fish Shellfish Immunol 25(4):341–350
4. Runft DL et al (2014) Zebrafish as a natural host model for Vibrio cholerae colonization and transmission. Appl Environ Microbiol 80(5):1710–1717
5. Sack DA et al (2004) Cholera. Lancet 363(9404):223–233
6. Senderovich Y, Izhaki I, Halpern M (2010) Fish as reservoirs and vectors of Vibrio cholerae. PLoS One 5(1):e8607
7. Traore O et al (2014) Occurrence of Vibrio cholerae in fish and water from a reservoir and a neighboring channel in Ouagadougou, Burkina Faso. J Infect Dev Ctries 8(10):1334–1338
8. Torres-Vitela MR et al (1997) Incidence of Vibrio cholerae in fresh fish and ceviche in Guadalajara, Mexico. J Food Prot 60(3): 237–241
9. Kiiyukia C et al (1992) Vibrio cholerae non-O1 isolated from ayu fish (Plecoglossus altivelis) in Japan. Appl Environ Microbiol 58(9): 3078–3082

Chapter 10

Transposon Sequencing of *Vibrio cholerae* in the Infant Rabbit Model of Cholera

Lauren M. Shull and Andrew Camilli

Abstract

Transposon sequencing, or Tn-seq, combines transposon mutagenesis and massively parallel sequencing to allow for rapid and high-throughput identification of genes that play roles in fitness within environments of interest. The bacterial pathogen *Vibrio cholerae* is an excellent candidate for Tn-seq screens due to the availability of a plasmid-based in vivo transposition system and the relative ease with which the cholera disease state can be modeled in animals. This chapter will describe a method for performing Tn-seq screens on *V. cholerae* in the infant rabbit model of cholera.

Key words Transposon sequencing, Tn-seq, *Vibrio cholerae*, Virulence, Cholera, Infant rabbit

1 Introduction

Tn-seq is a powerful tool to identify microbial genes that play important roles in a chosen selection condition. A transposon-mutagenized library is constructed such that each individual strain in the library harbors one transposon insertion at a random location in its genome, typically disrupting and inactivating a gene. Libraries of extremely high complexity can be easily generated so that, across the entire library, virtually every gene in the genome has been disrupted at multiple locations. Strains with insertions in essential genes drop out of the library during construction, and thus essential genes are putatively identified by the absence of insertions [1–3]. Genomic DNA from a portion of this starting library is isolated and retained for later PCR amplification and sequencing of the transposon junctions. The library is then subjected to some chosen selection, ideally one that is stringent and reproducible. The surviving bacteria are then recovered from the selection condition, outgrown if necessary, and genomic DNA is collected. The known sequence of the transposon is used as a forward PCR primer binding site, and a different known sequence is added downstream of the transposon by the method described

Aleksandra E. Sikora (ed.), *Vibrio Cholerae: Methods and Protocols*, Methods in Molecular Biology, vol. 1839, https://doi.org/10.1007/978-1-4939-8685-9_10,

below, providing a reverse PCR primer binding site. The transposon junctions are amplified by PCR, and the end of the transposon is used as a priming site for sequencing of the junctions [4]. The resulting reads are then mapped to the reference genome, identifying the location of each transposon insertion and relative abundance of each mutant in the library [1–3, 5, 6]. Mutants with insertions in genes required for the selection will be underrepresented in the output (or mutants in genes deleterious for the selection will be overrepresented), identifying such genes as candidates for validation and further study [1–3].

Although *V. cholerae* and other bacteria can be mutagenized with many types of transposon (e.g., Tn*5* or mariner-Himar1), construction of transposon insertion libraries in *V. cholerae* and other Gram-negative bacteria is facilitated by the availability of an in vivo transposition protocol mediated by the plasmid pDL1098 [5]. On the backbone of this plasmid are antibiotic resistance gene (*cat*; chloramphenicol resistance), mobilization origin of transfer (oriT), and a high-temperature-inducible, relaxed target-site specificity Tn*10* transposase (*tnpA*) [7]. Also on the plasmid is a mini-Tn*10* transposon (mTn*10*) harboring another antibiotic resistance gene (*aad9*; spectinomycin resistance). A second version of this plasmid was constructed, called pDL1093, which harbors *aphIII* encoding kanamycin resistance within the mTn*10*. In this system, replication of the plasmid and activity of the transposase are oppositely temperature-sensitive; at low temperatures (less than ~32 °C), the plasmid replicates, and the transposase are repressed, while at high temperatures (greater than ~38 °C), replication of the plasmid ceases, and the transposase gene is expressed. These temperature-sensitive controls allow maintenance of the plasmid without transposition at low temperature and with chloramphenicol selection. A simple shift of an early exponential phase broth culture to a high temperature and subsequent spectinomycin or kanamycin antibiotic selection selects for cells that have undergone a transposon insertion.

A library created in this manner can achieve a complexity of tens of thousands to hundreds of thousands of unique insertion sites in the genome [5]. A library of high complexity is in general desirable: statistical significance and confidence in results is more robust if multiple insertion site mutants of the same gene are analyzed. If only one or two insertion mutants of a particular gene are present in the input population, it is difficult to distinguish if under- or overrepresentation of that mutant in the output population is due to a change in fitness or to random fluctuations in the population. Conversely, if there are many insertions in the same gene and all or almost all behave similarly, it is reasonable to conclude that gene contributes to fitness in the selection.

Another important consideration in addition to library complexity is the bottleneck imposed by the selection. “Bottleneck” refers to the stochastic loss of a subset of the overall population, in contrast to the nonrandom loss of mutants with low fitness in the context of the selection. If the bottleneck is too constrictive, it is impossible to distinguish which mutants were unfit in the context of the selection from those that were lost on account of the bottleneck. This is especially significant during animal infection: while the infant rabbit model for cholera is supportive of highly complex libraries [8], the infant mouse model [9, 10] imposes a bottleneck such that only about 10^2–10^3 unique mutants can be recovered from an infection.

Infection efficiency can be improved (and thus the bottleneck effect minimized) in the infant rabbit model by sparing the inoculum from the highly acidic environment of the stomach. Pretreatment of the kits with the proton pump inhibitor Ranitidine reduces stomach acidity, and suspension of the inoculum in a sodium bicarbonate buffer further protects the bacteria from low pH [11].

An inoculum of 10^8–10^9 CFU prepared in 500 μL buffer is optimal for intragastric infection of infant rabbits. This dose results in cholera symptoms, namely, secretory diarrhea, within about 12 h after inoculation [11] and is also large enough that even a complex transposon library should be fully represented many times over. Animals are euthanized once symptomatic, at which point about 500 μL of cecal fluid can be recovered. Rabbits have a large cecum, and lumenal fluid collects there prior to release in the form of secretory diarrhea [11].

After a low-speed spin to pellet and remove clumps of mucus and host cells, the collected supernatant contains approximately 10^9 CFU/mL of planktonic, highly motile *V. cholerae*. This can be rapidly assessed by examining some of the collected supernatant by phase-contrast or dark-field microscopy. Because the number of *V. cholerae* recovered is at the lower limit for efficient genomic DNA isolation, outgrowth of the population is typically required. This is done in a rich broth medium supplemented with spectinomycin or kanamycin to limit growth of other bacteria. This outgrowth also allows for storage of backup cell pellets and cryopreserved glycerol stocks.

Libraries are prepared for sequencing by PCR-amplifying regions adjacent to transposon insertions, which can later be mapped to a reference genome to determine the location and frequency of each insertion [1–3, 5, 6]. Genomic DNA is first sheared by high-intensity sonication to lengths suitable for PCR amplification. Enzymatic methods (e.g., partial DNAse I digestion) may also be used to fragment the DNA. While the known transposon sequence can be used as the forward priming site for amplification, the random genomic sequence following the transposon provides no such reverse priming site. To circumvent this, a tail is appended to the 3′ ends of the sheared DNA using terminal deoxynucleotidyl transferase (TdT) and a 19:1 mixture of dCTP and the chain terminator ddCTP. This technique, called homopolymer tail-mediated ligation PCR, creates a poly-C tail approximately 20 nt in length as a recognition site for the reverse primer [4].

A second PCR is performed to barcode (index) the samples for Illumina multiplex sequencing using the product from the previous step as a template. Products from the second PCR step are mixed in the desired molar ratio, weighting the input sample most heavily, and this mixture is cleaned up and then ready for sequencing.

Tn-seq data is analyzed by mapping transposon junctions to the reference genome and comparing the number of reads at each insertion site in the input sample to the number in the output samples. Fitness scores are calculated for each insertion site and are then aggregated by gene [1–3]. Bioinformatic processing for Tn-seq analysis is explained in more detail in Subheading 3.4.

2 Materials

1. Water-cooled cuphorn sonicator.
2. Phosphate-buffered saline.
3. Luria-Bertani broth, Miller.
4. Antibiotics for selection: kanamycin, chloramphenicol.
5. 300 mM sodium bicarbonate buffer brought to pH 9.0 with 10 N NaOH (made fresh).
6. Ranitidine syrup 15 mg/mL: may require veterinary prescription.
7. Euthasol: requires veterinary prescription.
8. PE50 tubing: about 45 cm per unique inoculum, marked with a black Sharpie at 1, 2, 3, and 4.5 cm from the end, last 1–2 mm "softened" by pinching and pulling with fingernails; plus, one length of about 3 cm for administering Ranitidine.
9. 23 gauge needles and 1 mL syringes: one syringe + needle per unique inoculum, unmarked end of tubing pulled carefully onto needle and pulled to hilt; plus, two extra needles and syringes, one with 3 cm tubing pulled onto needle.
10. SYBR Safe DNA stain.
11. GelGreen DNA stain.
12. Edge Biosystems Performa DTR spin columns.
13. 19:1 dCTP:ddCTP, final concentration of 10 mM.
14. Easy-A cloning enzyme and buffer.
15. 10 mM dNTP mixture.
16. Terminal deoxynucleotidyl transferase (Promega) and buffer.

17. Genomic DNA extraction kit.
18. RNAse A, 10 mg/mL stock solution.
19. PCR cleanup kit.
20. Tn*10*-specific PCR primer 1 at 30 μM: 5′ GTGTGG GCACTCGACATATGACAAG 3′.
21. C-tail-specific primer 1 at 30 μM: 5′ GTGACTGGAGTTCA GACGTGTGCTCTTCCGATCTGGGGGGGGGGGG GGGG 3′.
22. Nested Tn*10*-specific primer 2 at 30 μM: this primer uses the Tn*10*-specific primer 1 sequence as a recognition site and introduces the P5 sequence (underlined) required for binding to the Illumina flow cell and cluster generation. 5′ AATGATACGGCGACCACCGAGATCTACACTCTTTGG GGGCCAAAATCATTAGGGGATTCATCAG 3′.
23. C-tail-specific primer 2 at 30 μM: this set of primers uses the C-tail primer 1 sequence as a recognition site and introduces the P7 sequence (underlined below) required for binding to the Illumina flow cell and cluster generation, an eight-base index that will be used to demultiplex libraries after sequencing, and a binding site for the Illumina Multiplexing Index Read Sequencing Primer (double underlined). One example of C-tail-specific primer 2 is 5′ CAAGCAGAAGACGGCATACGAGAT**AAAGGAAT**GT GACTGGAGTTCAGACGTGTGCTCTTCCGA TCT 3′ where the reverse complement of the index, 5′ ATTCCTTT 3′, is bolded.
24. Transposon junction sequencing primer at 30 μM: 5′ ACACTCTTTGGGGGCCAAAATCATTAGGGG ATTCATCAG 3′.
25. Illumina Multiplexing Index Read Sequencing Primer at 30 μM: 5′ GATCGGAAGAGCACACGTCTGAACTCCAGTCAC 3′.

3 Methods

3.1 Generation of a Transposon Library

1. Transform the in vivo transposition plasmid pDL1093 or pDL1098 into the target background strain by conjugation, electroporation, or natural transformation. The background strain must be able to grow at 30 °C and 40 °C and be sensitive to chloramphenicol and either kanamycin or spectinomycin. In *V. cholerae* and *Escherichia coli*, the plasmid can be maintained at 30 °C with chloramphenicol selection at 2.5 μg/mL. The mTn*10* confers either kanamycin resistance (pDL1093) or spectinomycin resistance (pDL1098), both at 100 μg/mL. In our experience, selection with kanamycin is more stringent

than with spectinomycin in *V. cholerae* and *E. coli*. Thus, the remainder of this protocol deals with pDL1093.

2. Grow the target strain with pDL1093 to stationary phase in Luria-Bertani Miller broth (LB) + 2.5 μg/mL chloramphenicol at 30 °C with aeration. Pre-warm 100 mL LB with 100 μg/mL kanamycin to 40 °C. Transfer 0.5 mL of the overnight culture to the pre-warmed LB, and grow with shaking at 40 °C overnight. At this high temperature, plasmid replication stops, and transposition is induced. Because transposition is inefficient, many of cells in the overnight culture will lack mTn*10* insertions and may or may not still harbor the plasmid. Therefore, a second passage is usually required (next step).
3. Pre-warm an additional 100 mL LB 100 μg/mL kanamycin to 40 °C, and transfer 100 μL of the overnight culture into this fresh medium. Grow once again to stationary phase at 40 °C (~6–8 h) to reduce the presence of cells that may still harbor the plasmid and/or that lack mTn*10* insertions. This culture is the final transposon library. A portion should be used to assay transposition frequency and presence of plasmid (**step 4**, below). In addition, the library should be frozen with 20% glycerol (v/v) at −70 °C in 1 mL aliquots for future use. It is prudent to store many (e.g., 50–100) single-use aliquots. Future amplification of the library to generate additional archived aliquots is not recommended due to skewing of strain representation.
4. To assay transposition efficiency and determine the percentage of cells in the final library that still harbor pDL1093, plate serial dilutions on LB agar plates grown at 30 °C (to enumerate all cells), plates with 100 μg/mL kanamycin grown at 40 °C (to enumerate cells with transposon insertions), and plates with both 2.5 μg/mL chloramphenicol and 100 μg/mL kanamycin grown at 30 °C (to enumerate cells still harboring the plasmid). A small percentage of cells still harboring the plasmid is tolerable as long as most of the cells in the library contain mTn*10* insertions.

3.2 Infection of 2- to 3-Day-Old Infant Rabbits and Recovery of Output Library

1. Thaw an aliquot of the *V. cholerae* mTn*10* library, and make several dilutions (e.g., 1/20 and 1/100) into 15 mL LB + 100 μg/mL kanamycin, and grow at 37 °C with aeration for several hours to stationary phase. Inoculate with a sufficient volume of the thawed library to ensure representation of the entire library. For a library of 10^5 complexity, i.e., 10^5 unique mTn*10* insertion strains, 10^8 CFU or 0.1 mL of the thawed library is sufficient to avoid a bottleneck. Dispose of the rest of the thawed aliquot properly; i.e., do not re-freeze thawed library glycerol stocks (*see* **Note 1**).

2. 3.5 h before the desired time of inoculation, remove the infant rabbits (kits) from their mother in a container tall enough to prevent them from escaping. Be sure to use a container that allows for airflow. All procedures with the rabbits require Institutional Animal Care and Use Committee approval.
3. Dilute Ranitidine syrup to 1.5 mg/mL with sterile water.
4. Number each kit on the back with a permanent marker and weigh the animal.
5. Using a 1 mL syringe fitted with a 23 gauge needle and about 3 cm of PE50 tubing such that at least 0.5 cm overhangs the end of the needle, administer 5 μg Ranitidine per gram of bodyweight directly into each kit's mouth 3 h before inoculation time (*see* **Note 2**). This is typically ~0.1 mL.
6. Place kits back in container, and keep at room-temperature (~24 °C) for 3 h, allowing for airflow. Do not place kits back with their dam, as she will often reject and may harm them.
7. Before infecting, prepare the bulk inoculum: pellet the library culture for 2 min at 8000 RCF, remove the supernatant, and resuspend cells to a final concentration of ~10^9 CFU/mL in freshly made sterile 300 mM sodium bicarbonate. Prepare a volume such that each kit can receive 0.5 mL of the inoculum with at least 4 mL extra. Pellet 1 mL of inoculum in duplicate, remove the supernatant, and store the cell pellets at −20 °C for later genomic DNA purification. In addition, temporarily store 1 mL of inoculum in 20% glycerol at −70 °C.
8. Inoculation of kits is ideally done with two people working together. Draw 1 mL of the inoculum into a 1 mL syringe. Place one of the 23 gauge needles with a long section of PE50 tubing onto the syringe (*see* Subheading 2, **step 8**). To account for the void volume in the tubing, place the end of the tube into the remaining inoculum, and depress the plunger just until fluid reaches the end of the tube. Keeping the end of the tubing submerged in the inoculum, draw the plunger back up to the 1 mL mark (*see* **Note 3**).
9. Intragastrically inoculate each kit: one person will insert the tubing into the mouth and gently feed more and more length down the esophagus, using the markings on the tube as a guide and typically inserting until the last marking is about even with the front teeth (*see* **Note 4**). There should not be significant resistance when inserting the tubing; if there is, withdraw the tubing, and try once more. Once the end of the tubing is in the stomach, the other person should slowly eject 500 μL. If backpressure is felt, the tubing may have kinked and increasing the pressure further could be dangerous in that it may cause the tubing to fly off the syringe and needle. Therefore, if backpressure is felt, withdraw the tubing from

the kit and insert once more. After inoculation, gently remove the tubing and inspect for any signs of trauma such as blood in the mouth or blood on the withdrawn tubing. If blood is observed, then the esophagus was likely damaged during the procedure and the animal must be euthanized, since bacteremia is likely.

10. Place animals back in the container; if more than one distinct inoculum is used, group animals by inoculum received to prevent contamination by secondary infection once the kits are symptomatic. Place at room temperature (~24 °C) with good airflow. Since the kits will become symptomatic after ~8 h, adsorbent bedding material should be placed at the bottom of the container.
11. Begin monitoring kits periodically for cholera symptoms and weight loss. Kits typically become symptomatic about 8–12 h postinoculation, but this varies with *V. cholerae* strain, inoculum size, and status of the kits. Kits should be euthanized after becoming symptomatic and before losing more than 10% of their starting body weight. Kits should also be euthanized if they stop moving freely (paresis) and/or are exhibit abnormal breathing (tachypnea).
12. Euthanize kits via CO_2 asphyxiation until unconscious, followed by intracardiac injection of 200 μg Euthasol per gram body weight using a 23 gauge needle and 1 mL syringe.
13. Sanitize the abdomen surface prior to dissection with 70% ethanol. Upon dissection of the abdomen, the cecum should be readily apparent on the animal's right side and swollen with fluid. The small intestine may also be swollen with fluid. Place a sterile container under the cecum (*see* **Note 5**), which has been gently pulled out to the side of the animal with forceps, and carefully cut open one side of cecum to allow liquid to drain into the container. Collect as much cecal fluid as possible, typically 500–800 μL, and transfer to a microcentrifuge tube.
14. Vortex the cecal fluid twice for 5 s to break up aggregates of bacteria. Reserve a small aliquot, 2–10 μL, of the cecal fluid prior to any processing, and use this to titer the bacterial burden by serial dilution and plating on LB agar + 100 μg/mL kanamycin (*see* **Note 6**).
15. Unprocessed cecal fluid does not contain enough bacteria to allow adequate genomic DNA recovery for sequencing and must therefore be outgrown. First, pellet mucus and eukaryotic cells by centrifugation for 2 min at 500 RCF. Transfer all of the supernatant to a new microcentrifuge tube. Outgrow in an appropriate volume of LB with 100 μg/mL kanamycin. After the culture has grown to stationary phase, store cell pellets and glycerol stocks as described above for the inoculum.

3.3 Library Preparation for Sequencing

1. Purify genomic DNA from individual cell pellets of each sample, including the inoculum (input) and each animal output. Each cell pellet should contain ~10^9 CFU. Include an RNAse A step to eliminate contaminating RNA. Measure the DNA concentration by spectrophotometry. Genomic DNA should be at a concentration of at least 50 ng/μL in a minimum volume of 50 μL. Check genomic DNA quality and quantity on a 0.8% agarose gel with SYBR Safe DNA stain.
2. Shear 1 μg of genomic DNA in a 50 μL volume in a PCR tube (or 2 μg in 100 μL in a parabolic-bottom 2 mL microcentrifuge tube) in a water-cooled cuphorn sonicator. Sonicate for 1 min at maximum amplitude, with a 5 s on/5 s off duty cycle. Pellet contents and repeat sonication.
3. Confirm shearing to a range of ~200–800 bp by running a small aliquot of each sample on a 2% agarose gel with GelGreen DNA stain. If necessary, repeat the sonication.
4. Remove small molecules from the sheared DNA by passing through an Edge Biosystems Performa DTR spin column (gel filtration). Quantify the concentration of DNA by spectrophotometry.
5. Append a poly-C tail to the sheared DNA by mixing 0.1 μg sheared DNA, 4 μL TdT reaction buffer, 1 μL 10 mM 19:1 dCTP/ddCTP, 0.5 μL TdT enzyme, and water to a volume of 20 μL. Incubate at 37 °C for 1 h, then heat-inactivate the enzyme at 75 °C for 20 min. Pass through an Edge Biosystems column once more.
6. PCR amplifies and indexes transposon-genomic junctions using 10 μL of the cleaned C-tailing reaction as template. Primers are listed in Table 1 below.
   (a) First PCR:

   10 μL C-tailed template.
   2 μL 10 mM dNTPs.
   5 μL Easy-A Cloning reaction buffer.
   1 μL mTn*10*-specific primer 1.
   3 μL C-tail-specific primer 1.
   1 μL Easy-A Cloning enzyme.
   Water to 50 μL.

   Begin the PCR program by heating at 95 °C for 2 min, then 25 cycles of 95 °C for 30 s, 58 °C for 30 s, and 72 °C for 2 min. Finish with another 2 min of extension time, and store at 10 °C or colder.

   (b) Second PCR:

   0.5 μL of first PCR.

   2 μL 10 mM dNTPs.

   5 μL Easy-A Cloning reaction buffer.

Table 1
Primers used in this protocol

Primer name	Sequence (5′–3′)	Reaction
Tn*10*-specific primer 1	GTGTGGGCACTCGACATATGACAAG	PCR 1
C-tail-specific primer 1	GTGACTGGAGTTCAGACGTGTGCTCTTCCGA TCT-GGGGGGGGGGGGGGGG	PCR 1
Nested Tn*10*-specific primer 2	AATGATACGGCGACCACCGAGATCTACACTC TTT-GGGGGCCAAAATCATTAGGGGATTCATCAG	PCR 2
C-tail-specific primer 2	CAAGCAGAAGACGGCATACGAGAT[N_8] GTGACTG-GAGTTCAGACGTGTGCTCTTCCGATCT	PCR 2
Transposon junction sequencing primer	ACACTCTTTGGGGGCCAAAATCATTAGGGGA TTC-ATCAG	Sequencing
Illumina Multiplexing Index Read Sequencing Primer	GATCGGAAGAGCACACGTCTGAACTCCAGTCAC	Sequencing

1 μL Nested mTn*10*-specific primer 2.

1 μL C-tail-specific primer 2 (use a different index ["N_8" in Table 1 below] for each sample).

1 μL Easy-A Cloning enzyme.

Water to 50 μL.

Begin the PCR program by heating at 95 °C for 2 min, then 18 cycles of 95 °C for 30 s, 52 °C for 30 s, and 72 °C for 2 min. Finish with a final extension of 2 min, and store at 10 °C or colder.

7. Visualize 2 and 8 μL of each reaction on a 2% agarose gel with GelGreen DNA stain. Load the 2 μL amounts in adjacent lanes, and the 8 μL amounts in adjacent lanes for later visual comparison. There should be a smear of products from about 200 to 1000 bp. Compare the different lanes by eye (*see* **Note** 7) to estimate the relative amounts of DNA in each: this information is used below for pooling at roughly equimolar amounts. An example gel analysis of HTML-PCR samples (5 μL was loaded) is shown in Fig. 1.
8. Pool a portion of the second PCR reactions to give an equimolar amount of each output and with the input at four times the molar concentration of each output. This ensures good coverage of the input, which is essential for analysis. Depending on library complexity, up to 24 libraries can be multiplexed in a single Illumina sequencing lane in this manner. Within a single

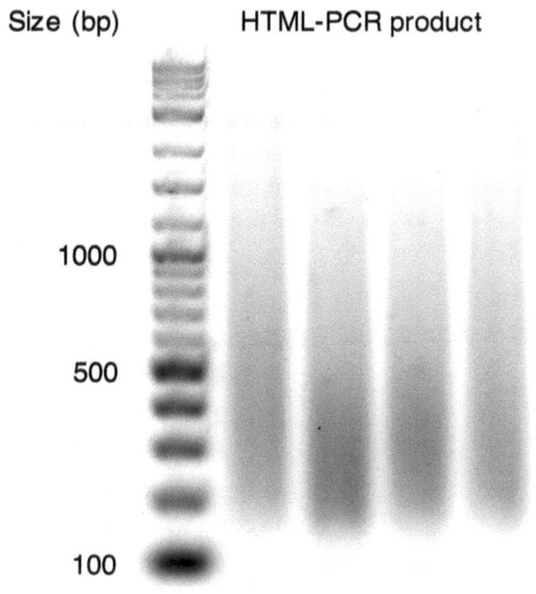

Fig. 1 An example of gel analysis of HTML-PCR samples

multiplexed mixture, each index primer must be unique. After pooling, temporarily store the remaining second PCR reaction products at −20 °C.

9. Clean the pooled sample with a PCR purification kit to remove residual primers and other small molecules. Quantify the concentration of DNA by spectrophotometry. Also visualize a small aliquot of the mixture on a 2% agarose gel with GelGreen DNA strain to ensure quality. The final DNA sample should be free of primers and primer-dimers, as these will interfere with sequencing. In some instances, gel purification of the 200–800 bp smear of products is required to remove primers and/or primer-dimers. Submit for Illumina single-end 50 cycle sequencing along with an aliquot of the custom transposon junction sequencing primer and information about the Illumina indexes used for multiplexing. You may also want to provide an aliquot of the Illumina Multiplexing Index Read Sequencing Primer; however, most Illumina sequencing core facilities have this primer.

3.4 Data Processing and Analysis

1. Map each file of demultiplexed sample reads to the reference genome using Bowtie. The poly-C tail will appear at the 3′ end of short junctional sequences and must be clipped bioinformatically for the reads to map properly.
2. For each sample, normalize the number of readers at each insertion site to the total number of reads from that sample.

3. Calculate fitness scores using the normalized mapped reads. The input sample is T1, and each output sample is T2. This calculation requires an expansion factor *d*, which represents the absolute value of the fold change in population size over the experiment, either calculated experimentally or estimated (*see* **Note 8**). The fitness calculation is as follows:

$$W_i = \frac{\ln\left(N_i(t_2) \times \frac{d}{N_i(t_1)}\right)}{\ln\left((1 - N_i(t_2)) \times \frac{d}{1 - N_i(t_1)}\right)}$$ where W_i is the fitness value

for insertion *I*, and $N_i(t_1)$ and $N_i(t_2)$ are the frequencies of the insertion at T1 and T2, respectively. Set the cutoff for number of reads per insertion at T1 to 15. This is necessary because the high sequencing error rate of the Illumina platform will generate a low number of reads that map to incorrect positions in the reference genome. In addition, set the program to report "zero" fitness scores, which are insertions that were present in the input sample but absent from the output sample.

4. Normalize fitness scores to known neutral genes. Using a list of 15–30 genes with known neutral fitness (e.g., defective genes, transposases, integron cassette genes but excluding toxin/antitoxin genes), calculate the multiplicative factor required to set the average fitness of all neutral genes in the fitness scores file to 1 (*see* **Note 8**). If this factor is itself not close to 1, then either the genes in the neutral list are not truly neutral or something went wrong during the selection such as a severe bottleneck. Normalize all the fitness scores by this factor. This step also determines the severity of the bottleneck by calculating the number of insertions in the neutral genes present in the input that are absent from the output.

5. Finally, aggregate insertional fitness scores by gene (*see* **Note 9**). Randomly exclude a proportion of zero fitness scores equal to the bottleneck calculation in **step 4** to correct for artificially low fitness scores due to the bottleneck. Calculate the mean fitness for each gene based on the fitness of all insertions in that gene. The aggregate fitness scores can be manipulated and sorted as desired, and statistics can be calculated if there are experimental replicates.

4 Notes

1. It is a good idea to start cultures at several dilutions to ensure the correct density at time of infection. Make sure the culture volume is sufficiently large for inoculation of animals, storing

of cell pellets for genomic DNA isolate, and freezing of one aliquot in glycerol.

2. Place toward the back of the mouth, and the kits will swallow the liquid. Too close to the front of the mouth and they will try to spit it out.
3. The same needle/tubing can be used for several animals; switching to a new tube is necessary only to prevent contamination when using a different inoculum or if the tubing becomes damaged.
4. If the animals are particularly small or large, this length may have to be altered to fit the animal; in this case, estimate the correct length of tubing to insert by holding a (clean) length of PE50 tubing next to the animal and marking as appropriate.
5. We use the lid of a petri plate, which is sterile and shallow enough to place below the cecum. Tilt the lid so that the fluid collects in one pool to facilitate transfer to a microcentrifuge tube.
6. If an estimate of the titer is needed before processing, observe a 5 μL aliquot by microscopy, and compare this to an in vitro-grown culture of a known titer. As a rule of thumb, cecal fluid from symptomatic kits typically contains about 10^9 *V. cholerae* per mL.
7. If desired, the molar amounts of DNA in each lane can be quantified by image analysis. Most gel imaging programs have quantification built in; the free software ImageJ also works well. Ensuring that no pixels are oversaturated, compare the DNA quantity in each lane to the known quantity of ladder loaded. Convert from mass to a molar quantity by estimating the peak size in base pairs of the smear and using the following formula:

$$\mu\text{g DNA} \times \frac{\text{pmol}}{660\ \text{pg}} \times \frac{10^6\ \text{pg}}{1\ \mu\text{g}} \times \frac{1}{\text{Length in base pairs}} = \text{pmol DNA}$$

8. In the case where there is not a neutral gene list available or the expansion factor is unknown, an alternative analysis protocol called HopCount may be used. The method for HopCount analysis is used and described in references 5 and 6.
9. Typically, we exclude insertions from the last 5% of the coding sequence, as these are less likely to truly “knock out” or inactivate the gene they are inserted into.

References

1. van Opijnen T, Bodi KL, Camilli A (2009) Tn-seq: high-throughput parallel sequencing for fitness and genetic interaction studies in microorganisms. Nat Methods 6:767–772. https://doi.org/10.1038/nmeth.1377
2. van Opijnen T, Camilli A (2013) Transposon insertion sequencing: a new tool for systems-level analysis of microorganisms. Nat Rev Microbiol 11:435–442. https://doi.org/10.1038/nrmicro3033
3. van Opijnen T, Lazinski DW, Camilli A (2014) Genome-wide fitness and genetic interactions determined by Tn-seq, a high-throughput massively parallel sequencing method for microorganisms. Curr Protoc Mol Biol 106:7.16.1–7.1624. https://doi.org/10.1002/0471142727.mb0716s106
4. Lazinski DW, Camilli A (2013) Homopolymer tail-mediated ligation PCR: a streamlined and highly efficient method for DNA cloning and library construction. BioTechniques 54:25–34. https://doi.org/10.2144/000113981
5. McDonough E, Lazinski DW, Camilli A (2014) Identification of in vivo regulators of the *Vibrio cholerae xds* gene using a high-throughput genetic selection. Mol Microbiol 92:302–315. https://doi.org/10.1111/mmi.12557
6. Klein BA, Tenorio EL, Lazinski DW et al (2012) Identification of essential genes of the periodontal pathogen Porphyromonas gingivalis. BMC Genomics 13:578. https://doi.org/10.1186/1471-2164-13-578
7. Bender J, Kleckner N (1992) Tn*10* insertion specificity is strongly dependent upon sequences immediately adjacent to the target-site consensus sequence. Proc Natl Acad Sci U S A 89:7996–8000
8. Kamp HD, Patimalla-Dipali B, Lazinski DW et al (2013) Gene fitness landscapes of *Vibrio cholerae* at important stages of its life cycle. PLoS Pathog 9:e1003800. https://doi.org/10.1371/journal.ppat.1003800
9. Ujiiye A, Nakatomi M, Utsunomiya A et al (1968) Experimental cholera in mice: I. First report on the oral infection. Trop Med 10:65–71
10. Klose KE (2000) The suckling mouse model of cholera. Trends Microbiol 8:189–191. https://doi.org/10.1016/S0966-842X(00)01721-2
11. Ritchie JM, Rui H, Bronson RT, Waldor MK (2010) Back to the future: studying cholera pathogenesis using infant rabbits. mBio 1:e00047-10. https://doi.org/10.1128/mBio.00047-10

Chapter 11

Isolation of Outer Membrane Vesicles Including Their Quantitative and Qualitative Analyses

Paul Kohl, Franz G. Zingl, Thomas O. Eichmann, and Stefan Schild

Abstract

Outer membrane vesicles (OMVs) are naturally secreted from the bacterial cell surface and therefore localized in the cell-free supernatant of bacterial cultures. Here we describe methods for crude and density gradient-purified OMV isolation and protocols for control analyses for protein profiling (SDS-PAGE), detection of indicator proteins (immunoblot analysis), lipid profiling (lipid extraction and LC-MS analysis), vesicle size determination (NanoSight), rough estimation of biomass (TrayCell™), as well as quantifications of defined OMV components, e.g., proteins (Bradford) and LPS (Purpald).

Key words OMVs, Ultracentrifugation, Density gradient purification, Quantification, NanoSight, TrayCell™, Bradford, Purpald, SDS-PAGE, Immunoblot, Lipid extraction, LC-MS

1 Introduction

Bacterial outer membrane vesicles are spherical, bilayered constrictions of the outer membrane of Gram-negative bacteria with 10–300 nm in size. As facsimiles of the outer membrane, these vesicles consist of outer membrane proteins, lipopolysaccharide (LPS), phospholipids, and periplasmic constituents [1].

In general, OMVs can be seen as transport vehicles of biological molecules, including lipids, LPS, enzymes, signaling molecules, toxins, and nucleic acids. In fact, they are the only secretion mechanism of Gram-negative bacteria capable to transport hydrophobic compounds. Several physiological roles for OMVs have been reported (e.g., inter- and intraspecies communication, nutrient predation, horizontal gene transfer, virulence, and immunomodulation) [2–5]. OMVs can stabilize proteins or serve as bioreactors as they bring enzymes in close proximity [6, 7]. In the case of *V. cholerae*, OMVs have been successfully characterized as vaccine candidates, which induce a robust protective immune response [8–10]. *V. cholerae* OMVs have been demonstrated to be internalized by human epithelial cells and induced inflammatory responses

Aleksandra E. Sikora (ed.), *Vibrio Cholerae: Methods and Protocols*, Methods in Molecular Biology, vol. 1839,
https://doi.org/10.1007/978-1-4939-8685-9_11,

via activation of NOD1 and NOD2 [11]. The vesicles contain several interesting effectors, including enzymes, like the proteases DegP and PrtV, and bacterial RNA which may affect the microbe-host interaction [12–14]. Not surprisingly, the research field on bacterial vesicles is growing exponentially.

Due to the diversified field of applications, numerous research groups try to isolate OMVs. The disparities in OMV isolation, even nomination of different OMV types after diverse altered isolation protocols, are tremendous and limit the interpretation and comparison of the results. Here we report an established isolation protocol for *V. cholerae* OMVs based on nearly a decade of experience [8, 15–19], which can be easily adjusted to other bacteria of interest.

Firstly, crude OMVs (cOMVs) are harvested by sequential filtration and centrifugation steps, which are then subjected to density gradient centrifugation to obtain purified OMVs (pOMVs). Furthermore, quantitative and qualitative assays are described to analyze biomass (TrayCell™), protein amount (Bradford), LPS amount (Purpald), vesicle diameters (NanoSight), the protein profile (SDS-PAGE) (*see* Fig. 1a), track specific proteins (immunoblot analysis) (*see* Fig. 1b), and the lipid profile (lipid extraction and LC-MS analysis (*see* Fig. 2) [8, 9, 11, 12, 15, 20, 21].

2 Materials

2.1 OMV Isolation

Unless stated otherwise, prepare all solutions with ddH_2O, and prepare and store all reagents at room temperature.

1. Components for liquid and solid media (herein: Luria-broth (LB)).
2. Petri dishes.
3. Autoclave.
4. Glassware: sterile bottles (1000 mL) and appropriate culture flasks (100 mL and 2000 mL).
5. Spectrophotometer (wavelength: 600 nm) with cuvette holder and cuvettes.
6. High-speed centrifuge: ideally refrigerated, including appropriate rotor with a loading capacity of at least 500 mL and autoclavable centrifuge tubes.
7. Bottle-top-filters 0.22 µm (low protein binding, we recommend polyethylensulfone).
8. Membrane vacuum pump.
9. Beckman ultracentrifuge Optima L-90K.
10. JA-10 Rotor, fixed angle, aluminum, 6 × 500 mL, max. speed 10,000 rpm, run at 17,700 × *g*.

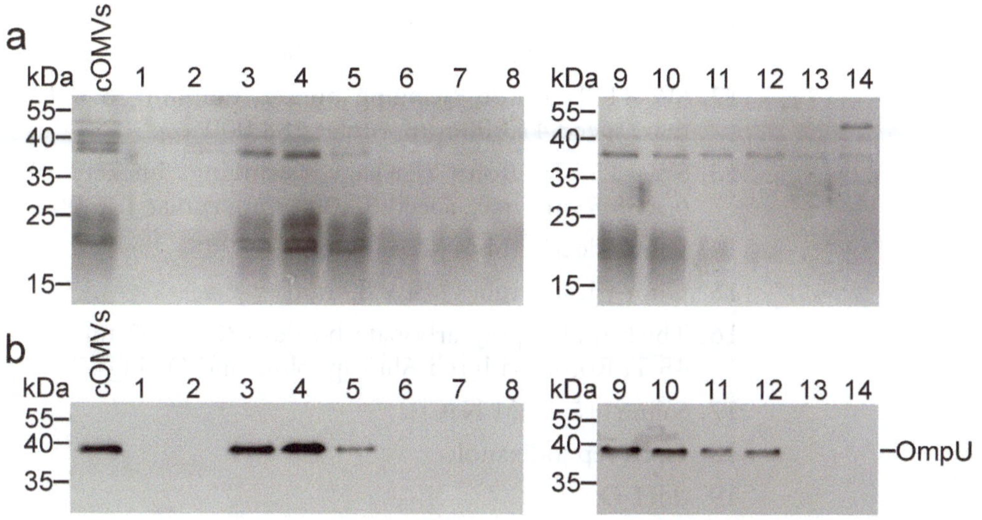

Fig. 1 OMV protein profile of *Vibrio cholerae* 01 El Tor. Depicted are the protein profiles (a) stained by Kang solution and immunoblot detecting the abundant outer membrane protein OmpU (b) of crude OMVs (lane 1, 7 μg protein equivalent loaded according to Bradford) and purified OMVs which were isolated out of 15 fractions taken after density gradient centrifugation (lanes 2–16, 6 μL of each fraction)

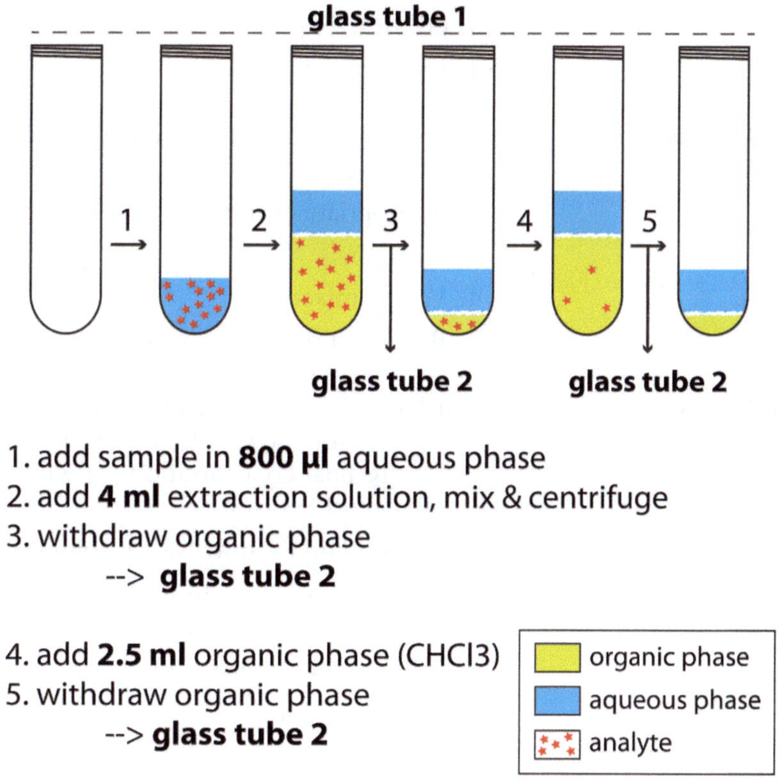

Fig. 2 Graphical illustration of the lipid extraction procedure. In course of the modified Folch extraction [21], neutral and unpolar compounds (e.g., lipids) are extracted into the organic phase of the biphasic system. Denaturated proteins and insoluble sample components form the interphase

11. Type 45 Ti Rotor, fixed angle, titanium, 6 × 94 mL, max. speed 45,000 rpm, run at 235,000 × *g*.
12. SW 41 Ti Rotor, swinging bucket, titanium, 6 × 13.2 mL, max. speed 41,000 rpm, run at 288,000 × *g*.
13. SW 32 Ti Rotor Package, swinging bucket, titanium, 6 × 38.5 mL, max. speed 32,000 rpm, run at 175,000 × *g*.
14. Ultra-Clear™ tubes (25 × 89 mm).
15. Ultra-Clear™ tubes (14 × 89 mm).
16. Thick-walled polycarbonate bottles (38 × 102 mm) for Type 45 Ti Rotor, with red Alu-cap, plug, and O-rings.
17. Saline (154 mM NaCl).
18. OptiPrep iodixanol.
19. ddH_2O.
20. Liquid nitrogen.

2.2 TrayCell™

1. TrayCell™ (Hellma).
2. Spectrophotometer with cuvette holder, applicable for TrayCell™ (wavelength: 260 nm and 280 nm).
3. ddH_2O.
4. Precision wipes.

2.3 Bradford

1. Spectrophotometer (wavelength: 595 nm) with cuvette holder and cuvettes.
2. Appropriate cuvettes.
3. Parafilm.
4. Vortex mixer.
5. Bio-Rad Protein Assay Dye Reagent Concentrate.

2.4 Purpald

1. KDO (3-deoxy-D-mannooctulosonic acid) standard: prepare samples for the standard curve in 50 μL aliquots with the following final concentrations of KDO: 0, 4, 8, 16, 20, 40, and 60 μg/mL.
2. 32 mM and 64 mM $NaIO_4$ solutions.
3. Purpald reagent: 136 mM 4-amino-3-hydrazino-5-mercapto-1,2,4-triazole dissolved in 2 N NaOH.
4. 2-Propanol (*see* **Note 1**).
5. Microtiter plates.
6. Microplate reader.

2.5 NanoSight

1. NanoSight LM10-HS instrument (Malvern Instruments).
2. Tracking analysis software.

3. Saline (154 mM NaCl), sterile filtered (0.02 μm).
4. Syringe.

2.6 SDS-PAGE

1. Separating gel (stock solution): 300 mL 2 M Tris/HCl, 16 mL 346 mM SDS, 84 mL ddH_2O, pH 8.8.
2. Stacking gel (stock solution): 100 mL 1 M Tris/HCl, 8 mL 346 mM SDS, 92 mL ddH_2O, pH 6.8.
3. Acrylamide/Bis-acrylamide (30%/0.8%, store at 4 °C).
4. 10% ammonium persulfate (APS, store at 4 °C).
5. *N,N,N',N'*-Tetramethylethan-1,2-diamine (TEMED, store at 4 °C).
6. Isopropanol.
7. 5× Laemmli buffer: 190 mM SDS, 70 mM EDTA, 47 mM $Na_2H_2PO_4$. 2 ddH_2O, 25% glycerol, 1.5 mM bromophenol blue, pH 7.2 (store at −20 °C).
8. 10× SDS-PAGE running buffer stock: 249 mM Tris/HCl, 18.8% glycine, 35 mM SDS, pH 8.3.
9. SDS-PAGE electrophoresis unit including power supply (e.g., Mini PROTEAN 3 System including gel casting stand, glass plates with 0.75 mm spacer, combs, and Power Pac 300).
10. Disposable lab ware: 50 mL conical tubes, 1.5 mL reaction tubes, paper towels.
11. Commercially available protein ladder (ideally prestained).
12. Staining solution by Kang: 111 mM aluminum sulfate hexadecahydrate, 10% EtOH, 2.34 μM coomassie blue G-250, 2% orthophosphoric acid [22].
13. Destaining solution: 10% ethanol (96%), 2% orthophosphoric acid (100%).
14. Gel-staining tray.
15. Image capturing system (e.g., ChemiDoc XRS system or similar).

2.7 Immunoblot

1. Nitrocellulose membranes.
2. Whatman paper.
3. CAPS buffer: 10 mM CAPS, 10% MeOH, pH 11.
4. TBS: 20 mM Tris/HCl, 150 mM NaCl, pH 7.5.
5. TBS-TT: 20 mM Tris/HCl, 250 mM NaCl, 0.05% Tween-20, 0.2% triton X-100, pH 7.5.
6. Blocking solution: 10% skim milk in TBS (store at 4 °C).
7. Appropriate primary antibody (herein: anti-OmpU generated in mouse).

8. Appropriate horse radish peroxidase (HRP)-conjugated secondary antibody (herein: anti-mouse IgG from goat).
9. Tank blotting system including power supply compatible with the SDS-PAGE unit.
10. Foil bag sealing device.
11. Gel-staining tray.
12. Laboratory rockers or shakers (ideally with three-dimensional motion).
13. Detection substrate for HRP (herein: Clarity™ Western ECL substrate).
14. Image capturing system (e.g., ChemiDoc XRS system or similar).

2.8 Lipid Extraction and Analysis

Unless stated otherwise, use high-purity solvents (HPLC grade or higher) to avoid unpolar contaminants and ddH_2O.

1. Solvents: chloroform ($CHCl_3$), methanol (MeOH), ddH_2O, acetic acid (CH_3COOH, glacial), 2-propanol (C_3H_8O), acetonitrile (C_2H_3N).
2. UPLC solvent modifier: ammonium acetate (1 M in ddH_2O), formic acid (conc.), phosphoric acid (8 mM in ddH_2O).
3. Leucine-enkephaline reference solution: 288 nM leucine-enkephaline acetate salt hydrate in ddH_2O/C_3H_3N, 1/1, v/v.
4. Glassware: 250 mL flask, 100 mL graduated cylinder.
5. Butylated hydroxytoluene solution (BHT): 500 μM in MeOH.
6. Internal standard mix (ISTD): 17:0-17:0 phosphatidylethanolamine, 34:0-PE, 17:0-17:0 phosphatidylcholine, 34:0-PC, 17:0 lyso-phosphatidylcholine, 17:0-LPC, 17:0-17:0-17:0 triacylglycerol, 51:0-TG, 2 mM in $CHCl_3$/MeOH, v/v.
7. Pyrex glass tubes (10 mL) and Teflon-sealed caps.
8. Laboratory rockers or shakers (ideally with three-dimensional motion).
9. Commercially available pipettes and tips (20, 200, 1000, 5000 μL).
10. Pasteur pipettes (15 cm) and Pasteur pipette holder.
11. HPLC vials and caps (300 μL with glass inlet, 1500 μL).
12. HPLC vial capper and decapper.
13. Reversed-phase column, BEH-C18, 2.1× 100 mm, 1.7 μm (Waters, Manchester, UK) or equivalent.
14. UPLC, Acquity UPLC system (waters) or equivalent technology.
15. Mass spectrometer, SYNAPT™G1 qTOF HD mass spectrometer (waters) equipped with an ESI source or equivalent technology.

3 Methods

3.1 Isolation of Crude OMVs (cOMVs)

1. Prepare sterile liquid LB broth media and LB agar plates for strain cultivation.
2. Streak strain which is stored on −80 °C on an LB agar plate, and incubate overnight at 37 °C.
3. Inoculate 25 mL LB broth in a 100 mL culture flask with a single colony from the agar plate, and incubate the culture flask for 16 h at 37 °C/180 rpm.
4. Inoculate 500 mL medium with 5 mL of the 25 mL culture in a 2000 mL culture flask for 8 h at 37 °C/180 rpm (*see* **Notes 2** and **3**).
5. After 8 h of incubation, measure the absorbance at 600 nm of the culture (*see* **Note 4**).
6. Fill/divide the 500 mL culture in centrifugation tubes, and centrifuge in a JA-10 Rotor in a high-speed centrifuge for 15 min at 10,000 × *g*/4 °C.
7. Sterile-filter the supernatant through a 0.22 μm bottle top filter, and plate at least 100 μL of the filtrate on an LB agar plate to confirm absence of living bacterial cells (*see* **Note 5**).
8. Load 70 mL of supernatant in thick-walled polycarbonate bottles, balance to 10^{-2} g accuracy, transfer them into a Type 45 Ti Rotor, and ultracentrifuge for 4 h at 150,000 × *g*/4 °C.
9. After ultracentrifugation immediately decant the supernatant, store tube briefly upside down to remove residual liquid, and resuspend the pellet in 70 μL saline. This results in a 1000-fold concentration of the OMVs compared to the original culture supernatant (*see* **Note 6**).
10. Proceed directly with the samples for further analyses, or store them at −20 °C (*see* **Note 7**).

3.2 Isolation of Density Gradient-Purified OMVs (pOMVs)

1. Fill an Ultra-Clear™ Tube (14 × 89 mm) from bottom to top with 1.4 mL layers of a 55, 50, 45, 40, 35, 30, and 25% OptiPrep iodixanol in ddH_2O. Mark the top of the 25% layer with a water-resistant marker (*see* **Note 8**).
2. Load on top of the density gradient up to 2 mL of OMV sample. The sample should contain approximately 1 mg protein determined by Bradford analysis. As a balance, it is required to run at least two density gradients at the same time.
3. Balance to 10^{-2} g accuracy, transfer them into a SW 41 Ti Rotor, and ultracentrifuge for 16 h at 150,000 × *g*/4 °C (*see* **Note 9**).
4. After ultracentrifugation remove the liquid until the marked line (start of the gradient), then proceed removing 700 μL

aliquots, and store each fraction separately (*see* **Note 10**). Fractions can be stored at −20 °C.

5. Transfer each 700 μL fraction in an Ultra-Clear™ Tube (25 × 89 mm), and fill the tube with saline.
6. Balance to 10^{-2} g accuracy, transfer them into a SW 32 Ti Rotor, and ultracentrifuge for 4 h at 150,000 × *g*/4 °C.
7. Resuspend the pellet with 200 μL saline and freeze at −20 °C (*see* **Note 7**).
8. Confirm presence of OMVs in each fraction by SDS-PAGE and immunoblot analysis (*see* Subheadings 3.7, 3.8, Fig. 1, and **Note 11**) using an abundant OMV marker protein (herein OmpU).

3.3 TrayCell™ Measurement (See Note 12)

1. Turn on spectrophotometer with single cuvette holder.
2. Choose a program for "nucleic acid analysis" or program the spectrophotometer to measure the absorbance at 260 and 280 nm.
3. Place TrayCell™ in spectrophotometer.
4. Wash TrayCell™ at least three times as follows: pipette 4 μL ddH_2O on TrayCell™, put the cap on, wait 1 min, then clean both TrayCell™, and cap carefully (*see* **Note 13**).
5. Place 4 μL of the sample on the TrayCell™ and put the cap on top. Measure the absorbance at 260 and 280 nm (UV) for each of the samples, and use as blank the same buffer which was used to dissolve the OMVs (saline). After each measurement wash at least once as described above (*see* **Note 14**).
6. After measurements wash again three times as described above.
   (a) Calculate the protein amount using the Warburg-Christian equation:

   $$\text{biomass of sample } [\mu g/\mu L] = (1.55 * E280 - 0.76 * E260) * \textit{cap-specific factor}.$$

7. Calculate the respective biomass content to 1 L original culture volume per OD unit (mg L^{-1} OD unit^{-1}) with the formula: $(b * 1000 * Vd)/(Vs * OD_{600})$.

   b: biomass amount [μg/μL]

   Vd: Volume sample is dissolved [mL]

   Vs: Volume supernatant in centrifuge tube [mL]

   OD_{600}: Absorbance 600 nm of the culture

3.4 Bradford Assays

1. Prepare appropriate dilutions of the OMV sample in saline with a final volume of 800 μL in standard cuvettes. We usually use 0.5, 2, and 5 μL.

2. Prepare for a calibration curve at least five dilutions of a protein standard (e.g., 100, 250, 500, 750, and 1500 µg/mL) in saline with a final volume of 800 µL in standard cuvettes. We use cOMVs as a standard which were measured three times by TrayCell™. If another in-house produced protein standard is used, the concentration should also be measured by TrayCell™ (*see* **Note 15**).
3. Add 200 µL Bio-Rad Protein Assay Dye Reagent Concentrate to each dilution of the protein standard as well as OMV sample. Use 200 µL Bio-Rad Protein Assay Dye Reagent Concentrate together with 800 µL saline as blank.
4. Immediately seal each cuvette with Parafilm and vortex.
5. Incubate samples for 30 min at room temperature.
6. Vortex samples, and measure the absorbance at 595 nm in a spectrophotometer.
7. Calculate the protein amount of each sample based on the regression curve of the protein standard.
8. In case of crude OMVs, you may calculate the OMV concentration (protein equivalent) present in the original bacterial culture per OD unit (mg L^{-1} OD $unit^{-1}$) with the formula: $(p * 1000 * Vd)/(Vs * OD_{600})$.

   p: protein amount [µg/µL]

   *V*d: Volume sample is dissolved [mL]

   *V*s: Volume supernatant in centrifuge tube [mL]

   OD_{600}: Absorbance 600 nm of the bacterial culture

3.5 Purpald Assays

1. Dilute OMV samples to be tested in ddH_2O. The dilution will depend mainly on the amount of LPS present in the samples; we usually start with 1:500, 1:1000, or 1:5000 (*see* **Note 16**).
2. Transfer 50 µL of each test dilution and KDO standard controls in duplicates with concentrations of 0, 4, 8, 16, 20, 40, and 60 µg/mL into a microtiter plate.
3. Add 50 µL 32 mM $NaIO_4$ to each well, mix by pipetting, and incubate for 25 min at room temperature.
4. Add 50 µL of Purpald reagent, mix by pipetting, and incubate for 20 min at room temperature (*see* **Note 17**).
5. Add 50 µL 64 mM $NaIO_4$ to stop reaction.
6. Add 20 µL 2-propanol to clear the solution of the foam that is formed in the previous step.
7. Measure the absorbance at 550 nm in a plate reader.
8. Calculate the lipooligosaccharide/lipopolysaccharide amount of each sample based on the regression curve of the KDO standard.

9. In case of crude OMVs, you may calculate the respective OMV concentration (lipooligosaccharide/lipopolysaccharide equivalent) present in the original bacterial culture per OD unit (mg L^{-1} OD $unit^{-1}$) with the formula: (l * 1000 * Vd)/(Vs * OD_{600}).

   l: lipooligosaccharide/lipopolysaccharide amount [μg/μL]

   Vd: Volume sample is dissolved [mL]

   Vs: Volume supernatant in centrifuge tube [mL]

   OD_{600}: Absorbance 600 nm of the culture

3.6 OMV Diameter Measurement by Nanoparticle Tracking Analysis

1. Dilute the OMV sample to a concentration of 0.1 μg/mL (protein equivalent measured by Bradford).
2. Apply a monochromatic laser beam at 488 nm to the OMV sample (*see* **Note 18**).
3. Take a video of 90 s at 23 °C.
4. Analyze particle movement by nanoparticle tracking analysis software with the minimal expected particle size, minimum track length, and blur setting all set to automatic.
5. Analyze video by nanoparticle tracking analysis software to determine the respective mean and mode OMV size.

3.7 OMV Protein Profile Analysis by Sodium Dodecyl Sulfate (SDS)-Polyacrylamide (PAA) Gel Electrophoresis (SDS-PAGE)

3.7.1 Preparation of Gels

1. Assemble Mini PROTEAN 3 system glass plates (0.75 mm spacer) in the casting stand (*see* **Note 19**).
2. Prepare a separating gel solution sufficient for four gels (15% PAA). Mix 5 mL separating gel from the stock solution, 4.9 mL ddH_2O, 10 mL Acrylamide/Bis-acrylamide (30%/0.8%), 50 μL 10% APS, and 30 μL 99% TEMED in a 50 mL conical tube.
3. Pour the separating gel solution immediately into the space between the two glass plates (*see* **Note 20**).
4. Gently overlay separating gel surface with isopropanol immediately.
5. Wait for 30 min to let the gel polymerize, and then remove the isopropanol (*see* **Note 21**).
6. Prepare a stacking gel solution sufficient for four gels. Mix 2.5 mL separating gel from the stock solution, 6.13 mL ddH_2O, 1.3 mL acrylamide/bis-acrylamide (30%/0.8%), 50 μL 10% APS, and 30 μL 99% TEMED in a 50 mL conical tube. Pour the stacking gel mixture on top of the polymerized separating gel, and insert the comb. Avoid creating bubbles. Wait for 30 min to let the gel polymerize.
7. Keep the gels moist at 4 °C overnight to ensure complete polymerization (*see* **Note 22**).

3.7.2 SDS-PAGE

1. Prepare the samples in a 1.5 mL reaction tube as follows: 3 μL 5× Laemmli buffer and 7 μg (protein equivalent measured by Bradford) of OMV sample, and fill up to a total volume of 15 μL with ddH_2O.
2. Boil the samples at 100 °C in a heat block for 10 min.
3. Prepare the SDS-PAGE electrophoresis unit: insert the gel and fill up with 1× SDS-PAGE running buffer.
4. Load the gel with the samples (*see* **Note 23**).
5. Apply 80 V for approx. 20 min until the dye front has migrated into the separating gel.
6. Pause the run and change the settings to 12 mA per gel inside the SDS-PAGE electrophoresis unit. Proceed the run until the dye front passes the bottom of the gel (*see* **Note 24**).
7. Turn off the power supply, and disassemble the SDS-PAGE electrophoresis unit. Carefully separate the gel from the glass plates using a spatula or a similar device. Remove the stacking gel, and place the residual gel in a staining tray (*see* **Note 25**).
8. Wash the gel with ddH_2O for 15 min.
9. Discard the ddH_2O. Stain the gel with Kang solution overnight for visualization of proteins or proceed with immunoblot analysis (*see* Subheading 3.8).
10. The next day discard the staining solution, and wash several times with ddH_2O (*see* **Note 26**).
11. Stained gels can be digitalized, e.g., by a molecular imager ChemiDoc XRS system (*see* Fig. 1a).

3.8 Immunoblot Analysis

1. For immunoblots, execute a SDS-PAGE (*see* Subheading 3.7, **steps 2–9**, and **Note 27**).
2. Cut out a nitrocellulose membrane and four Whatman papers slightly bigger than the gel (*see* **Note 28**).
3. Equilibrate the gel and the nitrocellulose membrane in CAPS buffer for 10 min.
4. Assemble the "sandwich": Fill CAPS buffer inside a suitable plastic container and insert the grid, one sponge, two Whatman papers, the SDS-gel, the nitrocellulose membrane, and two Whatman papers in exactly that order. Use a fixed roller on the Whatman paper, rolling back and forth to get rid of all air bubbles (*see* **Note 29**).
5. Add another sponge on the top, close the grid, and insert it into an electro-blotting tank filled with CAPS buffer (*see* **Note 30**).
6. Transfer the proteins at 220 mA for 90 min.

7. Disassemble the sandwich, put the membrane in a gel-staining tray, and wash it two times in TBS for 10 min each, on a rocker at room temperature (*see* **Note 31**).
8. Incubate with 10% skim milk in TBS over night at 4 °C on a rocker.
9. Wash membrane once with TBS.
10. Dilute 0.8 μL of the primary antibody (Anti-OmpU) in 4 mL of 10% milk in TBS (1:5000 dilution), and mix briefly by vortexing.
11. Seal the membrane and the milk-antibody solution in disposal bags using a foil bag sealing device (*see* **Note 32**).
12. Incubate sealed membrane for 2 h on a rocker at room temperature.
13. Remove the membrane from the bag, place it into a clean gel-staining tray, and wash it twice with TBS-TT.
14. Wash membrane once with TBS.
15. Dilute 1 μL of the secondary antibody in 7.5 mL of 10% skim milk in TBS (1:7500 dilution), and mix briefly by vortexing.
16. Incubate membrane with secondary antibody for 1 h on a rocker at room temperature.
17. Wash membrane four times with TBS-TT.
18. Wash membrane once with TBS.
19. Prepare a Clarity™ Western ECL substrate solution (1 mL peroxide solution + 1 mL Luminol/enhancer solution), and incubate membrane with the solution for 5 min (*see* **Note 33**).
20. Expose blot to an X-ray film, or detect luminescence by a molecular imaging device (e.g., ChemiDoc XRS System) (Fig. 1b).

3.9 Lipid Extraction and Analysis

3.9.1 Preparations

1. Purge glassware with $CHCl_3$ prior to usage (dishwasher clean Pyrex glass tubes, graduated cylinder, flasks).
2. Prepare 4 mL $CHCl_3$/MeOH/CH_3COOH/BHT/ISTD (2/1/1%/500 nM/1 μM) per sample in a glass flask (extraction solution, modified Folch et al.) [21] (*see* **Note 34**).
3. Prepare 2.5 mL $CHCl_3$ per sample in a glass flask (*see* **Note 34**).

3.9.2 Lipid Extraction (See Fig. 2)

Pyrex Glass Tube 1

1. Add OMV preparations in a total volume of 800 μL ddH_2O or buffer.
2. Add 4 mL of extraction solution, and mix samples for 60 min at room temperature.
3. Centrifuge the samples for 20 min at 1000 × *g*/room temperature to achieve phase separation.

4. Withdraw organic (lower) phase, and transfer to Pyrex glass tube 2 (*see* **Note 35**).
5. Add 2.5 mL $CHCl_3$ and mix samples for 20 min on room temperature.
6. Centrifuge the samples 20 min at 1000 × *g*/room temperature to achieve phase separation.
7. Withdraw organic (lower) phase, and transfer to Pyrex glass tube 2 (*see* **Note 35**).

Pyrex Glass Tube 2

1. Evaporate the combined organic phase under nitrogen stream (N_2).
2. Resolve extracted lipid analytes in $C_3H_8O/CHCl_3/MeOH$ (7/2/1, v/v/v) for LC-MS analysis (*see* **Note 36**).
3. Transfer resolved lipid analytes to a HPLC vial, and use it for LC-MS analysis or storage at −20 °C.

3.9.3 LC-MS Analysis (Modified Knittelfelder et al. [20])

LC Parameters and Chromatographic Separation

1. Prepare solvent A, $MeOH/ddH_2O$ (1/1, v/v) and solvent B, C_3H_8O both containing 10 mM ammonium acetate, 0.1% formic acid, and 8 μM phosphoric acid as modifier.
2. Install reversed-phase column and heat column compartment to 50 °C.
3. Cool sample manager to 8 °C and insert sample vials.
4. Set flow to 55% solvent A at 300 μL/min flow.
5. Inject 5 μL or 10 μL of sample for positive or negative ionization mode, respectively.
6. Run samples using a linear gradient.
   (a) The gradient starts at 55% solvent A and decreases to 10% solvent A within 11 min.
   (b) Within the following 2 min, solvent A percentage decreases to 0%.
   (c) Starting conditions (55% solvent A) are restored within 0.5 min, and the column is re-equilibrated for 1.5 min, resulting in a total run time of 15 min.

Mass Spectrometer Parameters and Method

1. Set capillary temperature to 100 °C, desolvation temperature to 400 °C, and activate N_2 as nebulizer gas.
2. Set capillary voltage to 2.6 kV in positive and 2.1 kV in negative ionization mode.
3. Define two alternating scan modes for MS^E analysis.
   (a) First mode: Set mass range to *m/z* 50–1800, and scan time to 0.3 s and data collection to centroid.

(b) Second mode: Set mass range to *m/z* 50–1800, scan time to 0.3 s, data collection to centroid, and apply a collision energy ramp from 10 to 60 V to fragment generated ions.

4. Apply a lock spray containing leucine-enkephaline ($[M^{+}H]^{+}$: *m/z* 556.2771, $[M^{-}H]^{-}$: *m/z* 554.2615) via an external pump (0.2 mL/min, 1/13 split) for continuous mass correction.
5. Set lock mass measurement interval to 15 s in the MS method.
6. Acquire the data with MassLynx software (waters), and analyze "lipids of interest" using high-resolution mass and MS^{E} fragmentation (e.g., PE, positive mode: $[M^{+}H]^{+}$ and *m/z* 141 neutral loss, negative mode $[M^{-}H]^{-}$, and specific fatty acids fragments as $RCOO^{-}$). Use ISTD to correct for extraction and ionization variations.

4 Notes

1. These solutions should be prepared fresh each time.
2. 500 mL is the minimum amount to cultivate because the capacity of the ultracentrifuge is 420 mL of supernatant (six times 70 mL using the Type 45 Ti Rotor). Approximately 10% of the culture volume is lost during the bacterial cell removal.
3. The exact cultivation period needs to be empirically determined. We recommend isolating vesicles from late-log phase cultures as a relatively high, reproducible yield without too much contamination by lysed cells can be obtained. In case of the *V. cholerae* isolate used herein, late-log phase is reached after 8 h of incubation.
4. The absorbance at 600 nm is a measure of cell density and important for calculation of the OMV concentration in the original culture volume normalized to OD (e.g., mg L^{-1} OD $unit^{-1}$).
5. The supernatant can be stored in the fridge but should be centrifuged within 24 h to avoid degradation of the OMVs. Dependent on the subsequent assays protease inhibitors can be added.
6. OMVs detach quickly from the pellet on the bottom of the tube. A huge amount of OMVs can be lost when too much time passes before decanting the supernatant. Due to the shape of the centrifuge tubes, not all the liquid can be decanted; storing the tubes upside down avoids residual liquid coming into contact with the pellet containing the OMVs. The last drops of liquid on the inside of the tube can be wiped off with a clean towel. In case of low yield, the resuspension volume should be reduced.

7. The storage largely depends on their further use. Once isolated the OMVs are very stable. One might consider snap freezing with liquid nitrogen and storing them at −80 °C to increase stability.
8. Avoid vibrations of the tube when pouring the gradient. The layers composed of different OptiPrep iodixanol concentrations must not be mixed. We find it best to pipette the first 400 μL of each layer with a 200 μL pipette. Afterward, one might change to the 1 mL pipette for the remaining 1 mL of the layer.
9. Use a slow to modest acceleration/deceleration setting to avoid to undesirable vibrations of the ultracentrifuge.
10. For a comprehensive analysis, even smaller volumes of each fraction can be taken. Please consider that this will increase the amount of fractions, and each has to be ultracentrifuged again independently.
11. There are two different populations of OMVs in *V. cholerae*, which can be separated by density gradient centrifugation.
12. The TrayCell™ method is a fast and easy technique that requires only a small amount of the sample. However, complex compositions of proteins, nucleic acid, lipids, etc. cannot be correctly analyzed by this method. In addition, OMV samples do not represent a homogenous solution. Due to these limitations, the TrayCell™ should only be used to get a rough estimation of the biomass in the sample. We have also tried the NanoDrop technology, but TrayCell™ results have proven more reliable and reproducible. It should be noted that the method relies on absorption at 260 and 280 nm. Do not measure OMV samples, which are contaminated with substance, that have a measurable absorption within this wavelength range. For example, pure OMV samples still contain residual OptiPrep iodixanol, which has a decent absorption at 260 nm and 280 nm. Thus, the biomass calculations of pure OMVs by TrayCell™ will be incorrect.
13. Clean only with precision wipes to avoid scratches on the glass surface of the TrayCell™.
14. It is not necessary to remove the TrayCell™ out of the cuvette holder between measurements. Avoid direct contact between pipette tip and surface of the TrayCell™ to avoid scratches on the glass surface.
15. Choosing the optimal protein sample for the standard curve is critical. The Bradford assays relies on aromatic residues in the protein samples. OMVs naturally contain large amounts of hydrophobic membrane proteins. Thus, the frequently used BSA standard curve will not allow correct protein calculations.

Preferably, use cOMVs of the same strain or purified membrane proteins (e.g., outer membrane porins).

16. Please note that some plastic lab ware interferes with the Purpald reagent; the best option is to use low-binding reaction tubes. We suggest testing the available material by running a standard curve including a blank to avoid unnecessary waste of OMV samples.
17. A yellow color immediately appears but vanishes very quickly. The final purple color intensifies over time.
18. Each frame should contain 10–100 detectable particles. Use appropriate dilutions of the OMV sample.
19. Clean the surfaces of the glass to remove all dirt and dust particles, which would otherwise interfere in protein band migration or staining and thereby reduce the quality.
20. 15% polyacrylamide gels resolve best between 20 and 50 kDa, which matches the molecular weight of the most abundant outer membrane proteins of *V. cholerae*. Leave sufficient space for the stacking gel. The border between stacking and separation gel should be approx. 0.5 cm below the lowest point of the comb.
21. Isopropanol can later easily be removed by soaking Whatman paper strips, which are carefully placed between the glass plates. Polymerization can be visually checked by the unused separating gel solution still inside the 50 mL conical tube.
22. Gels can be stored at 4 °C for several days, but extended storage has adverse effects on the resolution of the bands in the gel. Keep the gels moist by wrapping them in wet paper tissues or plastic wrap.
23. One slot should always be loaded with a prestained protein standard. Empty slots should be filled with Laemmli buffer to avoid a "smiley" effect on the gel.
24. The end of the run should be adjusted to obtain the best separation at the molecular weight range of interest. Use the bands of the prestained protein standard as a guideline.
25. The gel will stick to one of the two glass plates. Submerge the glass with the attached gel upside down in the staining tray filled with ddH_2O. The gel will detach from the glass and go smoothly into the plastic container.
26. If necessary, gels can be destained 10–60 min with a destaining solution [10% ethanol (96%) and 2% orthophosphoric acid (100%)]. Afterward, wash gels again with ddH_2O.
27. We recommend running one additional gel in parallel and subjecting it to Kang staining as described in Subheading 3.7, **step 2**. This gel serves as a loading control.

28. Make the membrane and Whatman pieces approx. 0.5 cm larger than the gel. A paperboard dummy of the right size shortens the preparation time.
29. The CAPS buffer should cover the entire sandwich composed of membrane, gel, and Whatman paper. Air bubbles can prevent protein transfer.
30. Due to alterations in transfer efficiency and antibody binding, an immunoblot is considered a semiquantitative method.
31. Try to avoid pouring solutions directly onto the membrane.
32. Make sure, before sealing the last part of the bag, to get rid of all air bubbles. This way fewer antibodies are consumed.
33. Shake staining tray, including membrane, gently by hand to ensure complete coverage of the membrane with detection solution.
34. Prepare 10% more of the extraction solution considering pipetting error.
35. Leave 10–15% of the organic phase in Pyrex glass tube 1 to avoid aqueous contaminations.
36. Resolve lipids according to 0.66 μg OMV protein/μL solvent (protein of the extracted sample) to yield a concentration suitable for direct MS measurement.

Acknowledgments

This work was supported by the Austrian FWF grants P25691 to S.S. and W901-B12 (DK Molecular Enzymology) to P.K., F.Z., and S.S. as well as BioTechMed Graz (Flagship project "Secretome").

References

1. Beveridge TJ (1999) Structures of gram-negative cell walls and their derived membrane vesicles. J Bacteriol 181(16):4725–4733
2. Grenier D, Mayrand D (1987) Functional characterization of extracellular vesicles produced by *Bacteroides gingivalis*. Infect Immun 55(1):111–117
3. Gankema H, Wensink J, Guinee PA, Jansen WH, Witholt B (1980) Some characteristics of the outer membrane material released by growing enterotoxigenic *Escherichia coli*. Infect Immun 29(2):704–713
4. McBroom AJ, Kuehn MJ (2007) Release of outer membrane vesicles by Gram-negative bacteria is a novel envelope stress response. Mol Microbiol 63(2):545–558
5. Schooling SR, Beveridge TJ (2006) Membrane vesicles: an overlooked component of the matrices of biofilms. J Bacteriol 188(16):5945–5957
6. Alves NJ, Turner KB, Walper SA (2016) Directed protein packaging within outer membrane vesicles from *Escherichia coli*: design, production and purification. J Vis Exp (117). https://doi.org/10.3791/54458
7. Alves NJ, Turner KB, Daniele MA, Oh E, Medintz IL, Walper SA (2015) Bacterial nanobioreactors—directing enzyme packaging into bacterial outer membrane vesicles. ACS Appl Mater Interfaces 7(44):24963–24972. https://doi.org/10.1021/acsami.5b08811

8. Leitner DR, Lichtenegger S, Temel P, Zingl FG, Ratzberger D, Roier S, Schild-Prüfert K, Feichter S, Reidl J, Schild S (2015) A combined vaccine approach against *Vibrio cholerae* and ETEC based on outer membrane vesicles. Front Microbiol 6:823. https://doi.org/10.3389/fmicb.2015.00823
9. Leitner DR, Feichter S, Schild-Prufert K, Rechberger GN, Reidl J, Schild S (2013) Lipopolysaccharide modifications of a cholera vaccine candidate based on outer membrane vesicles reduce endotoxicity and reveal the major protective antigen. Infect Immun 81(7):2379–2393. https://doi.org/10.1128/IAI.01382-12
10. Schild S, Nelson EJ, Bishop AL, Camilli A (2009) Characterization of *Vibrio cholerae* outer membrane vesicles as a candidate vaccine for cholera. Infect Immun 77(1):472–484
11. Bielig H, Rompikuntal PK, Dongre M, Zurek B, Lindmark B, Ramstedt M, Wai SN, Kufer TA (2011) NOD-like receptor activation by outer membrane vesicles from *Vibrio cholerae* non-O1 non-O139 strains is modulated by the quorum-sensing regulator HapR. Infect Immun 79(4):1418–1427. https://doi.org/10.1128/IAI.00754-10
12. Rompikuntal PK, Vdovikova S, Duperthuy M, Johnson TL, Ahlund M, Lundmark R, Oscarsson J, Sandkvist M, Uhlin BE, Wai SN (2015) Outer membrane vesicle-mediated export of processed PrtV protease from *Vibrio cholerae*. PLoS One 10(7):e0134098. https://doi.org/10.1371/journal.pone.0134098
13. Sjostrom AE, Sandblad L, Uhlin BE, Wai SN (2015) Membrane vesicle-mediated release of bacterial RNA. Sci Rep 5:15329. https://doi.org/10.1038/srep15329
14. Altindis E, Cozzi R, Di Palo B, Necchi F, Mishra RP, Fontana MR, Soriani M, Bagnoli F, Maione D, Grandi G, Liberatori S (2015) Protectome analysis: a new selective bioinformatics tool for bacterial vaccine candidate discovery. Mol Cell Proteomics 14(2):418–429. https://doi.org/10.1074/mcp.M114.039362
15. Schild S, Nelson EJ, Camilli A (2008) Immunization with *Vibrio cholerae* outer membrane vesicles induces protective immunity in mice. Infect Immun 76(10):4554–4563
16. Roier S, Leitner DR, Iwashkiw J, Schild-Prufert K, Feldman MF, Krohne G, Reidl J, Schild S (2012) Intranasal immunization with Nontypeable *Haemophilus influenzae* outer membrane vesicles induces cross-protective immunity in mice. PLoS One 7(8):e42664
17. Roier S, Fenninger JC, Leitner DR, Rechberger GN, Reidl J, Schild S (2013) Immunogenicity of *Pasteurella multocida* and *Mannheimia haemolytica* outer membrane vesicles. Int J Med Microbiol 303(5):247–256. https://doi.org/10.1016/j.ijmm.2013.05.001
18. Roier S, Blume T, Klug L, Wagner GE, Elhenawy W, Zangger K, Prassl R, Reidl J, Daum G, Feldman MF, Schild S (2015) A basis for vaccine development: comparative characterization of *Haemophilus influenzae* outer membrane vesicles. Int J Med Microbiol 305(3):298–309. https://doi.org/10.1016/j.ijmm.2014.12.005
19. Roier S, Zingl FG, Cakar F, Durakovic S, Kohl P, Eichmann TO, Klug L, Gadermaier B, Weinzerl K, Prassl R, Lass A, Daum G, Reidl J, Feldman MF, Schild S (2016) A novel mechanism for the biogenesis of outer membrane vesicles in Gram-negative bacteria. Nat Commun 7:10515. https://doi.org/10.1038/ncomms10515
20. Knittelfelder OL, Weberhofer BP, Eichmann TO, Kohlwein SD, Rechberger GN (2014) A versatile ultra-high performance LC-MS method for lipid profiling. J Chromatogr B Anal Technol Biomed Life Sci 951–952:119–128. https://doi.org/10.1016/j.jchromb.2014.01.011
21. Folch J, Lees M, Sloane Stanley GH (1957) A simple method for the isolation and purification of total lipides from animal tissues. J Biol Chem 226(1):497–509
22. Kang D, Gho YS, Suh M, Kang C (2002) Highly sensitive and fast protein detection with coomassie brilliant blue in sodium dodecyl sulfate-polyacrylamide gel electrophoresis. Bull Kor Chem Soc 23(11):1511–1512

Chapter 12

Utilization of *Vibrio cholerae* as a Model Organism to Screen Natural Product Libraries for Identification of New Antibiotics

Aleksandra E. Sikora, Richard Tehan, and Kerry McPhail

Abstract

The development of antibiotic-resistant bacteria requires increasing research efforts in drug discovery. *Vibrio cholerae* can be utilized as a model gram-negative enteric pathogen in high- and medium-throughput screening campaigns to identify antimicrobials with different modes of action. In this chapter, we describe methods for the optimal growth of *V. cholerae* in 384-well plates, preparation of suitable microtiter natural product sample libraries, as well as their screening using measurements of bacterial density and activity of type II secretion-dependent protease as readouts. Concomitant LC-MS/MS profiling and spectral data networking of assay sample libraries facilitate dereplication of putative known and/or nuisance compounds and efficient prioritization of samples containing putative new natural products for further investigation.

Key words Type II secretion system, Serine proteases, Cell-based assay, Natural products, High-throughput screening, LC-MS

1 Introduction

Treating multidrug-resistant microbes requires seeking new agents that will either kill bacteria or target pathways contributing to their virulence. There are many advantages to this latter approach, but particularly attractive is the ability to selectively eradicate infectious agents while preserving the human bacterial microbiome. The survival and virulence of many gram-negative bacteria are associated with their capability to produce and secrete toxins and degradative enzymes including proteases, lipases, chitinases, and neuraminidases via the type II secretion (T2S) pathway. The T2S system is widely distributed and well conserved among bacteria isolated from different ecological niches (reviewed in: [1]). Bacteria that utilize T2S systems cause diseases in humans, animals, and plants. In humans, these microbes are often the etiological agents of life-threatening diseases such as cholera, other diarrheal diseases, and pneumonia, including Legionnaires' disease [1–6]. Therefore, the

Aleksandra E. Sikora (ed.), *Vibrio Cholerae: Methods and Protocols*, Methods in Molecular Biology, vol. 1839, https://doi.org/10.1007/978-1-4939-8685-9_12, © Springer Science+Business Media, LLC, part of Springer Nature 2018

T2S system is an attractive target for developing new antimicrobial agents. Depending on the bacterial species, the T2S machinery is responsible for secretion from 1 (in *Klebsiella*) to more than 20 (*Vibrio cholerae*, *Legionella pneumophila*) exoproteins [1–6]. As several of these secreted proteins contribute to disease, and bacteria defective in the T2S are highly attenuated in virulence, secretion through the T2S machinery is considered as one of the major virulence mechanisms. The T2S system transports proteins through the bacterial outer membrane and releases them in the host. This process is mediated by distinct interactions between 12 and 15 different T2S components that constitute a nanomachine spanning the entire bacterial cell envelope [1–3]. In *V. cholerae*, the causative agent of cholera, the T2S system is responsible for extracellular transport of over 20 proteins including multimeric cholera toxin, which is mainly responsible for the symptoms of the disease [5]. Our studies have shown that genetic inactivation of the T2S system in *V. cholerae* results in pleiotropic effects, including loss of secretion, cell envelope instability, induction of several stress responses, and increased sensitivity of the bacteria to antimicrobial agents [7–9]. Moreover, *V. cholerae*, lacking a functional T2S system, is unable to survive the passage through the gastrointestinal tract of infant mice, the cholera model for infection [7]. Together, these findings indicate that the T2S system is an attractive target for developing new antimicrobial agents and suggest that chemical compounds that are able to disrupt one of the many interactions within the T2S system may be used to block the secretion process and prevent manifestation of the disease. Recently high-throughput assays have been developed to target the T2S in a plant pathogen *Dickeya dadantii* [10] and a serious nosocomial human pathogen, *Acinetobacter baumannii* [11].

Natural product extracts are widely recognized to be important sources of new small molecules for the discovery and production of potential pharmaceuticals, especially at a time when pharmaceutical pipelines of new efficacious agents remain critically low [12]. Efficient "mining" of available repositories of natural products to assess their biological activity for development of a potential therapeutic lead or molecular tool is critical. Preparation of pure biologically active metabolites remains resource-intensive, and a majority are only accessible from the crude extracts of plants, invertebrates, and microorganisms in sub- to low-milligram quantities [13, 14]. Thus, functional microtiter assays are essential for testing available high-value pure natural products and also for targeted purification of the putative active components from natural products extracts that are assay hits. Unfortunately, crude extracts are typically highly complex mixtures containing a large number and diversity of natural products and present many challenges for high-throughput biological testing, such as heterogeneous solubility and/or non-specific cytotoxicity [15]. Preliminary fractionation of natural product extracts, using column chromatography or solid phase extraction to generate unique fractions of varying polarity, allows natural products to be screened more efficiently and facilitates subsequent targeted purification of the active fractions [16, 17]. This "prefractionation" is reported to be a beneficial technique for effective evaluation of natural product extracts used in high-throughput screening applications in that up to 80% or more of hits generated from HTS of fractions were inactive in crude form [16, 18].

The availability of HTS assays compatible with more complex mixtures and demand for more dynamic context-dependent (e.g., ecological) comparative studies of NP production has prompted survey-type analyses of patterns of molecular diversity, rather than characterization of one component at a time. The inherent sensitivity of mass spectrometry (MS) and dramatic advances in techniques [19], such as atmospheric ionization, means that workflows using liquid or gas chromatography hyphenated with MS can be designed to evaluate the molecular complexity of large numbers of variably complex samples in sub- to low-microgram amounts. In a given sample, a variety of molecules can be tracked in situ with precision and accuracy that provides detailed structure information, especially when real-time fragmentation of observed mass ions (MS/MS or MS^n) is incorporated. This LC-MS/MS profiling provides information on HPLC retention time, UV light absorption, putative molecular or pseudomolecular ion, and mass fragmentation pattern. Thus, integrating LC-MS/MS profiling with biological assay data for samples may enhance the utilization of HTS screening toward clinical lead or target development [13].

This chapter describes assays that can be used for detection and assay-guided isolation of new antimicrobials acting either as traditional antibiotics or inhibitors of the T2S pathway in *V. cholerae*. The assay has been evaluated and shown to be amenable to screening natural product extracts, fractions, and pure compounds generated from field-collected marine and terrestrial cyanobacteria, terrestrial fungi, and laboratory cultures of deep-sea hydrothermal vent-associated organisms.

1.1 High-Throughput Screening

We describe techniques to grow *V. cholerae* under standard laboratory conditions in Luria-Bertani media at 37 °C in 384-well plates for optimal bacterial growth and assessment of functionality of the T2S system in an HTS format. This assay utilizes a quantitative measurement of the activity of T2S-dependent serine proteases VesA, VesB, and VesC [9]. Protease VesB contributes to about 80% of the activity that is measured using a highly sensitive synthetic fluorogenic substrate, N-*tert*-butoxycarbonyl-Gln-Ala-Arg-7-amino-4-methylcoumarin (Boc-QAR-AMC), commonly used for monitoring proteolytic activity of trypsin-like proteases [20]. A protease recognizes and cleaves a peptide bond between Arg and AMC, resulting in the release of a fluorescent molecule, which can be quantitatively measured (Fig. 1a, [7, 9, 21, 22]). The serine protease activity can be detected in the

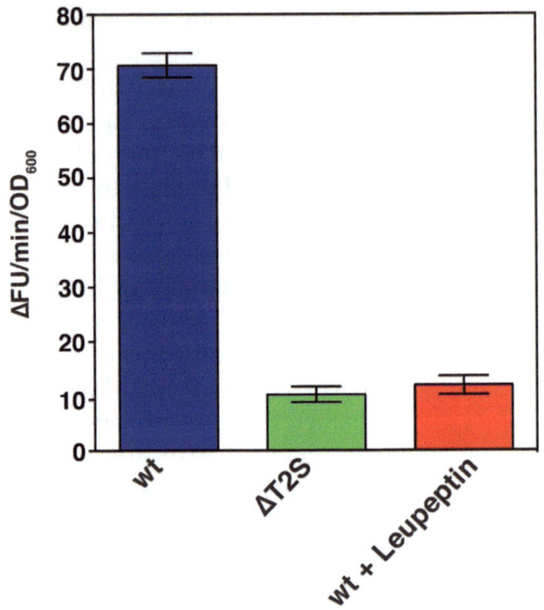

Fig. 1 T2S-dependent serine protease activity. Protease activity was assessed in stationary-phase cultures of *V. cholerae* N16961 wild type grown in the presence of DMSO (blue) and isogenic ΔT2S mutant (green) and wild type grown in the presence of leupeptin (red) using the proteolytic substrate *N*-tert-butoxy-carbonyl-Gln-Ala-Arg-7-amido-4-methylcoumarin. The rate of hydrolysis (relative fluorescence units [FU]/min/OD_{600} unit) is presented as the means of three independent experiments ± SEM

medium of growing bacterial cultures and is about sixfold higher in wild-type bacteria than the T2S-deficient strain (Fig. 1). To design an assay amenable for a high-throughput format, the number of experimental manipulations should be minimized. Therefore, we recommend using as a positive control, *V. cholerae* grown in the presence of leupeptin, an inhibitor of VesB protease [22]. The activity of the protease in the presence of the inhibitor is as low as in the cultures of T2S mutants (Fig. 1). Wild-type *V. cholerae* grown in media supplemented with the solvent vehicle for test samples (DMSO) displays a robust protease activity and is utilized as a negative control for the assay (Fig. 1). The assay is outlined in Fig. 2.

1.2 Natural Product Samples

Natural product samples for testing in the assay come predominantly from 96-well plates of 10% aqueous DMSO stock solutions (10 mg/mL) that represent diverse *organic* extracts, partly purified fractions (from column chromatography, solid phase extraction and HPLC), and pure compounds. The isolation and purification protocols used to generate these samples are biased toward a medium polarity suited to DMSO dissolution. The hygroscopic nature of DMSO can lead to concentration changes of the stored microtiter stock solutions over relatively short periods; 10% water in DMSO is recommended as a storage solvent, when solubility

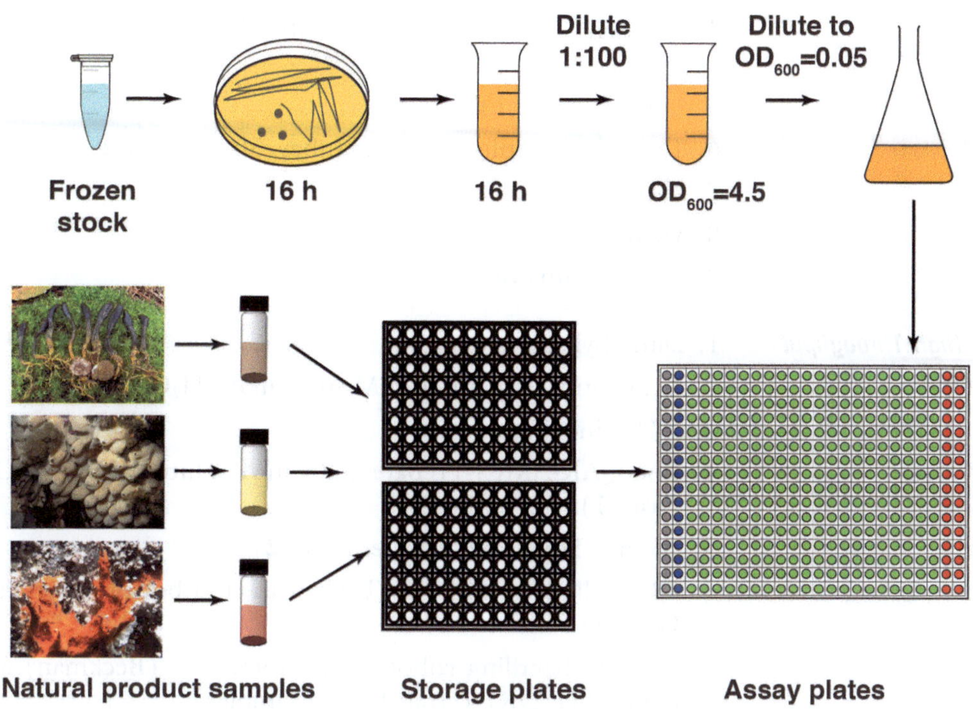

Fig. 2 Experimental outline of preparation for HTS campaign utilizing *V. cholerae* as a model organism. Preparation for HTS campaign includes isolation of natural products from different sources, their extraction, and their fractionation and aliquoting into microtiter storage plates followed by delivery at desired concentrations into sample library plates. *V. cholerae* is streaked out from frozen stock on LB agar plate, a single colony is used to inoculate LB medium, and bacteria are cultured under standard conditions for 16 h. Subsequently, the overnight culture is diluted 1:100 into fresh LB broth and grown until OD_{600} of ~4.5 is reached. At this point, bacterial culture is diluted again to an OD_{600} of 0.05, and bacteria are delivered to individual wells of microtiter screening plates using plate dispenser

allows, since additional absorption of atmospheric water is negligible [23].

Concomitant LC-MS/MS profiling of plated natural product stock solutions enhances the high-throughput biological screening by permitting some or all of the following: (1) preliminary assessment of putative active component (e.g., a mass peak common to neighboring active samples) and (2) possible dereplication of assay hits as known compounds, which could be known structures with known antibiotic activity, known structures not reported previously as antibiotic, or new antibiotic structures.

2 Materials

2.1 Growth of V. cholerae from a Frozen Stock and in Liquid Medium

1. *V. cholerae* frozen stock (*see* Subheading 2.1 in Chapter 1).
2. Luria-Bertani (LB) agar plates.
3. Ice bucket with cover: fill the bucket with ice before taking out bacterial stock from the 80 °C freezer.

4. Wooden applicator stick, toothpick, or inoculating loop, sterile.
5. Luria-Bertani (LB) broth.
6. Test tubes, sterile or conical tubes, 50 mL, sterile.
7. 250 mL sterile flask.
8. Vortex.
9. 37 °C incubator.

2.2 High-Throughput Screening

1. Dimethyl sulfoxide.
2. Leupeptin: prepare 10 mM stock in dd H_2O.
3. 37 °C incubator.
4. Boc-QAR-AMC: Prepare 0.25 mM solution in DMSO (*see* **Note 1**).
5. 25 mM Hepes pH 7.5 (*see* **Note 2**).
6. 384-well plates with lids (black with clear bottom, sterile; e.g., Corning 3712).
7. A liquid-handling robot [e.g., Biomek FX (Beckman) with a pin tool (HDR) to transfer test samples].
8. A plate dispenser (e.g., BioTek).
9. A multimode microplate reader to monitor absorbance and fluorescence (e.g., BioTek).
10. Plastic containers and plastic wrap (e.g., Saran wrap) to create humid chambers (*see* **Note 3**).

2.3 Natural Product Test Samples

1. Dimethyl sulfoxide (DMSO, ACS grade).
2. Microtiter deep-well (1.64 mL) plates for sample library storage (Axygen, P-DW-20-C).
3. Crude organic extracts, fractions, and pure natural product compounds (*see* **Note 4** for source and specific sample preparation).
4. Aliquots of natural product samples from point 3 in microtiter sample library plates: prepare 10 mg/mL stock solutions of crude extracts and fractions and 2 mM solutions of pure compounds in 10% H_2O-DMSO.
5. Aliquots of natural product samples for LC-MS/MS profiling: estimate an appropriate sample mass for injection considering complexity of samples and given a target injection volume of 10 μL per LC-MS run (e.g., 1 mg/mL crude extracts and fractions and 100 μg/mL pure compounds in 4:1 MeOH-H_2O or other suitable solvent).
6. Prepare a vial of 100% MeOH as a solvent blank.

3 Procedures

3.1 Growth of *V. cholerae* from a Frozen Stock and in Liquid Medium (Outlined in Fig. 2).

1. Using a sterile inoculating loop or wooden stick, remove a small amount of frozen *V. cholerae* stock from −80 °C freezer (*see* **Note 5**).
2. Streak onto LB agar to isolate single colonies.
3. Incubate for 16 h in incubator set to 37 °C.
4. The following day, using aseptic technique, prepare the starter culture by inoculating a single *V. cholerae* colony into 5 mL of LB in sterile test tube or 50 mL conical tube (*see* **Note 6**).
5. Cover the test tube or conical tube, and incubate bacteria with shaking for 16 h at 37 °C.
6. Back dilute the culture (1:100) in 30 mL of fresh LB medium in 250 mL sterile flask.
7. Grow bacteria with shaking at 37 °C until stationary phase of growth (OD_{600} about 4.5; *see* **Note 7**).
8. Finally, dilute this culture to OD_{600} of 0.05 in fresh LB, pre-warmed to 37 °C (*see* **Note 8**).

3.2 High-Throughput Screening

HTS of small-molecule and natural product libraries is routinely performed in an HTS facility, which, in addition to providing the libraries of stock plates, is equipped with state-of-the-art liquid handlers and multimode microplate readers that can be integrated for automated HTS applications. However, the screening procedures can be also executed on a smaller scale in a standard laboratory by using 96-well microtiter plates, plate dispenser, a microplate reader, and multichannel manual pipettes (Fig. 2).

1. Use the multidrop dispenser to deliver 20 μL of LB medium to each well of rows 1–22 on the 384-well plates (black wells and clear bottom).
2. After loading all plates, use the microplate dispenser to add 20 μL of LB medium containing protease inhibitor leupeptin and DMSO (at 10 μM and 0.5% final concentration, respectively) to the last two rows on individual plates.
3. Wash the microplate dispenser with excess of LB medium to remove any traces of leupeptin and to prime the tubing and cassette.
4. Transfer the test samples (0.2 μL/well) from library stock plates to the assay plates using a liquid-handling robot with a pin tool (*see* **Note 9**).
5. Add 20 μL of the culture from Subheading 3.1, **step 8**, into each well of the 384-well plate with the exception of the first row (LB alone).

6. Incubate all plates for 16 h at 37 °C in a humid chamber (*see* **Note 10**).
7. The following day, remove microtiter plates from the humid chamber, and place them on a bench with lids slightly open to remove any condensation.
8. Carefully transfer the microtiter plates on to a plate stacker, and examine bacterial growth by measuring the turbidity of the cultures at OD_{600} using the plate reader.
9. Prepare the solution of 0.25 mM Boc-QAR-AMC in 25 mM Hepes pH 7.5, and add 10 μL to each well in the 384-well microtiter plate using the plate dispenser; immediately shake each plate for 30 s to mix all reagents.
10. In the meantime, set a timer to monitor the time from addition of the protease substrate, Boc-QAR-AMC, and after 10 min proceed to measuring VesB activity in an end-point assay at excitation and emission wavelengths of 380 nm and 450 nm, respectively, using microplate reader (*see* **Note 11**).

3.3 LC-MS/MS Screening

1. Develop appropriate HPLC and MS methods for the chemotypes of sample known or predicted to be present (*see* **Note 12**).
2. Acquire data for the samples in random order, performing a methanol wash run between each sample run.
3. Develop an LC-MS data processing and analysis workflow for NP dereplication and identification (*see* **Note 13**).

4 Notes

1. Prepare a 0.25 mM solution of Boc-QAR-AMC in DMSO. The peptide can be purchased from Bachem, Peptides International, or other company. Store the substrate at −20 °C or for longer periods of time at −80 °C. The final concentration used in the assay is 50 μM.
2. The final concentration of Hepes in the assay is 5 mM.
3. The final concentration of leupeptin is 10 μM in LB.
4. Natural product extracts were prepared from field-collected marine and terrestrial Panamanian *Cyanobacteria*, South African tunicates, laboratory cultures of terrestrial *Tolypocladium* fungi, and laboratory cultures of (deep-sea hydrothermal vent) DSV-associated microorganisms. Cyanobacterial field collections and DSV cultures were extracted and pre-fractionated as described previously [10]. Deep-frozen collections of South African tunicates were lyophilized before extraction with 2:1 dichloromethane methanol (DCM-MeOH), and the resulting concentrated

organic extracts were pre-fractionated by size exclusion chromatography on Sephadex LH20 in 1:3 DCM-MeOH, followed by RP_{18} SPE or flash chromatography using a MeOH-H_2O solvent gradient (for instance, as described by Sikorska et al. [24]). Laboratory cultures of *Tolypocladium* fungi were grown on top of cellophane on SM (minimal) agar medium in 6-well plates for 7 days. The harvested cell mass was extracted in 1:1 MeOH-H_2O (extract A) followed by 2:1 DCM-MeOH (extract B) to provide aqueous and organic extracts for each culture, respectively, which were subjected to RP_{18} SPE (100 mg cartridges). Aqueous extracts (A) were eluted from the SPE cartridge in 1:1 MeOH-H_2O and 100% MeOH, while organic extracts (B) were eluted from the SPE cartridge in 1:1 MeOH-H_2O and 100% MeOH followed by a DCM wash.

5. Place the frozen bacterial stock in a bucket filled with ice to prevent thawing. Thawing will decrease bacterial viability over time.
6. For optimal aeration, we recommend using 50 mL sterile conical tubes.
7. Under these conditions, it takes about 4 h to reach an OD_{600} of 4.5.
8. The volume of this culture depends on the number of 384-well plates used for each screening procedure. Calculate the amount by multiplying 20 μL by the number of wells on each plate and the total number of microtiter plates. Importantly, add further amount of LB to include the dead volume for filling the tubing in the plate dispenser.
9. Do not deliver the test samples to the first two and last two columns on each plate as they will be used as a background and positive control, respectively.
10. A humid chamber can be prepared in a plastic container holding a beaker filled with water. After placing microtiter plates in the container, covering it with a generous amount of plastic wrap creates a humid chamber.
11. The first column on each plate is used as a blank for both the absorbance and fluorescence readings, the second column is used as a negative control (*V. cholerae* with no compound added), and the two last columns are utilized as a positive control for the assay.
12. For profiling extracts and first-tier fractions of *Cyanobacteria* and tunicates, we used an AB SCIEX 3200 Q-TRAP LC-MS/MS equipped with Shimadzu dual LC20 HPLC. For the HPLC method, we used a Phenomenex Synergi 4 μ Hydro-RP, 250 × 4.60 mm column, and linear gradient of (20:80 to 100:0) MeCN + 0.1% formic acid in H_2O + 0.1% formic acid. For the MS method, we used an information-dependent acquisition (IDA) method with enhanced MS (EMS) survey

scan from m/z = 100–1500 and five dependent enhanced product ion (EPI) scans of the most abundant ion peaks in m/z range 200–1700.

13. Several metabolomics tools are available online for LC-MS/MS data processing and analysis [25, 26]. We used Global Natural Products Social Molecular Networking (GNPS), which is an open-access platform for analysis and dereplication of MS/MS data using crowd-sourced MS/MS spectral libraries [27]. High-resolution MS/MS data are most suitable for rigorous NP dereplication and structure elucidation.

Acknowledgments

This work was funded by the start-up funds to A.E.S. and an Oregon Sea Grant under award number NA10OAR4170059 (project number R/BT-48) from the National Oceanic and Atmospheric Administration's National Sea Grant College Program to K.L.M. and A.E.S.

References

1. Sandkvist M (2001) Type II secretion and pathogenesis. Infect Immun 69(6):3523–3535
2. Sandkvist M (2001) Biology of type II secretion. Mol Microbiol 40(2):271–283
3. Cianciotto NP (2005) Type II secretion: a protein secretion system for all seasons. Trends Microbiol 13(12):581–588. https://doi.org/10.1016/j.tim.2005.09.005
4. Cianciotto NP (2009) Many substrates and functions of type II secretion: lessons learned from Legionella pneumophila. Future Microbiol 4:797–805. https://doi.org/10.2217/fmb.09.53
5. Sikora AE (2013) Proteins secreted via the type II secretion system: smart strategies of Vibrio cholerae to maintain fitness in different ecological niches. PLoS Pathog 9(2):e1003126. https://doi.org/10.1371/journal.ppat.1003126
6. Johnson TL, Waack U, Smith S, Mobley H, Sandkvist M (2015) Acinetobacter baumannii is dependent on the type II secretion system and its substrate LipA for lipid utilization and in vivo fitness. J Bacteriol 198(4):711–719. https://doi.org/10.1128/JB.00622-15
7. Sikora AE, Lybarger SR, Sandkvist M (2007) Compromised outer membrane integrity in Vibrio cholerae Type II secretion mutants. J Bacteriol 189(23):8484–8495. https://doi.org/10.1128/JB.00583-07
8. Sikora AE, Beyhan S, Bagdasarian M, Yildiz FH, Sandkvist M (2009) Cell envelope perturbation induces oxidative stress and changes in iron homeostasis in Vibrio cholerae. J Bacteriol 191(17):5398–5408. https://doi.org/10.1128/JB.00092-09
9. Sikora AE, Zielke RA, Lawrence DA, Andrews PC, Sandkvist M (2011) Proteomic analysis of the Vibrio cholerae type II secretome reveals new proteins, including three related serine proteases. J Biol Chem 286(19):16555–16566. https://doi.org/10.1074/jbc.M110.211078
10. Tran N, Zielke RA, Vining OB, Azevedo MD, Armstrong DJ, Banowetz GM, McPhail KL, Sikora AE (2013) Development of a quantitative assay amenable for high-throughput screening to target the type II secretion system for new treatments against plant-pathogenic bacteria. J Biomol Screen 18(8):921–929. https://doi.org/10.1177/1087057113485426

11. Waack U, Johnson TL, Chedid K, Xi C, Simmons LA, Mobley HLT, Sandkvist M (2017) Targeting the Type II secretion system: development, optimization, and validation of a high-throughput screen for the identification of small molecule inhibitors. Front Cell Infect Microbiol 7:380. https://doi.org/10.3389/fcimb.2017.00380
12. Newman DJ, Cragg GM (2016) Natural products as sources of new drugs from 1981 to 2014. J Nat Prod 79(3):629–661. https://doi.org/10.1021/acs.jnatprod.5b01055
13. Kurita KL, Linington RG (2015) Connecting phenotype and chemotype: high-content discovery strategies for natural products research. J Nat Prod 78(3):587–596. https://doi.org/10.1021/acs.jnatprod.5b00017
14. Espen H, Jeanette HA (2016) Screening for marine natural products with potential as chemotherapeutics for acute myeloid leukemia. Curr Pharm Biotechnol 17(1):71–77. https://doi.org/10.2174/1389201016666150817095537
15. Henrich CJ, Beutler JA (2013) Matching the power of high throughput screening to the chemical diversity of natural products. Nat Prod Rep 30(10):1284–1298. https://doi.org/10.1039/c3np70052f
16. Wagenaar M (2008) Pre-fractionated microbial samples—the second generation natural products library at Wyeth. Molecules 13(6):1406
17. Tu Y, Jeffries C, Ruan H, Nelson C, Smithson D, Shelat AA, Brown KM, Li X-C, Hester JP, Smillie T, Khan IA, Walker L, Guy K, Yan B (2010) Automated high-throughput system to fractionate plant natural products for drug discovery. J Nat Prod 73(4):751–754. https://doi.org/10.1021/np9007359
18. Henrich CJ, Cartner LK, Wilson JA, Fuller RW, Rizzo AE, Reilly KM, McMahon JB, Gustafson KR (2015) Deguelins, natural product modulators of NF1-defective astrocytoma cell growth identified by high-throughput screening of partially purified natural product extracts. J Nat Prod 78(11):2776–2781. https://doi.org/10.1021/acs.jnatprod.5b00753
19. Bouslimani A, Sanchez LM, Garg N, Dorrestein PC (2014) Mass spectrometry of natural products: current, emerging and future technologies. Nat Prod Rep 31(6):718–729. https://doi.org/10.1039/C4NP00044G
20. Kawabata S, Miura T, Morita T, Kato H, Fujikawa K, Iwanaga S, Takada K, Kimura T, Sakakibara S (1988) Highly sensitive peptide-4-methylcoumaryl-7-amide substrates for blood-clotting proteases and trypsin. Eur J Biochem 172(1):17–25
21. Lybarger SR, Johnson TL, Gray MD, Sikora AE, Sandkvist M (2009) Docking and assembly of the type II secretion complex of Vibrio cholerae. J Bacteriol 191(9):3149–3161. https://doi.org/10.1128/JB.01701-08
22. Gadwal S, Korotkov KV, Delarosa JR, Hol WG, Sandkvist M (2014) Functional and structural characterization of Vibrio cholerae extracellular serine protease B, VesB. J Biol Chem 289(12):8288–8298. https://doi.org/10.1074/jbc.M113.525261
23. Schopfer U, Engeloch C, Stanek J, Girod M, Schuffenhauer A, Jacoby E, Acklin P (2005) The Novartis compound archive—from concept to reality. Comb Chem High Throughput Screen 8(6):513–519. https://doi.org/10.2174/1386207054867328
24. Sikorska J, Hau AM, Anklin C, Parker-Nance S, Davies-Coleman MT, Ishmael JE, McPhail KL (2012) Mandelalides A–D, cytotoxic macrolides from a new Lissoclinum species of South African tunicate. J Org Chem 77(14):6066–6075. https://doi.org/10.1021/jo3008622
25. Sugimoto M, Kawakami M, Robert M, Soga T, Tomita M (2012) Bioinformatics tools for mass spectroscopy-based metabolomic data processing and analysis. Curr Bioinforma 7(1):96–108. https://doi.org/10.2174/157489312799304431
26. Yi L, Dong N, Yun Y, Deng B, Ren D, Liu S, Liang Y (2016) Chemometric methods in data processing of mass spectrometry-based metabolomics: a review. Anal Chim Acta 914:17–34. https://doi.org/10.1016/j.aca.2016.02.001
27. Wang M, Carver JJ, Phelan VV, Sanchez LM, Garg N, Peng Y, Nguyen DD, Watrous J, Kapono CA, Luzzatto-Knaan T, Porto C, Bouslimani A, Melnik AV, Meehan MJ, Liu W-T, Crusemann M, Boudreau PD, Esquenazi E, Sandoval-Calderon M, Kersten RD, Pace LA, Quinn RA, Duncan KR, Hsu C-C, Floros DJ, Gavilan RG, Kleigrewe K, Northen T, Dutton RJ, Parrot D, Carlson EE, Aigle B, Michelsen CF, Jelsbak L, Sohlenkamp C, Pevzner P, Edlund A, McLean J, Piel J, Murphy BT, Gerwick L, Liaw C-C, Yang Y-L, Humpf H-U, Maansson M, Keyzers RA, Sims AC, Johnson AR, Sidebottom AM, Sedio BE, Klitgaard A, Larson CB, Boya PCA, Torres-Mendoza D, Gonzalez DJ, Silva DB, Marques LM, Demarque DP, Pociute E, O'Neill EC, Briand E, Helfrich EJN, Granatosky EA, Glukhov E, Ryffel F, Houson H, Mohimani H, Kharbush JJ, Zeng Y, Vorholt JA, Kurita KL, Charusanti P, McPhail KL, Nielsen KF, Vuong L, Elfeki M, Traxler MF, Engene N, Koyama N, Vining OB, Baric R, Silva RR, Mascuch SJ,

Tomasi S, Jenkins S, Macherla V, Hoffman T, Agarwal V, Williams PG, Dai J, Neupane R, Gurr J, Rodriguez AMC, Lamsa A, Zhang C, Dorrestein K, Duggan BM, Almaliti J, Allard P-M, Phapale P, Nothias L-F, Alexandrov T, Litaudon M, Wolfender J-L, Kyle JE, Metz TO, Peryea T, Nguyen D-T, VanLeer D, Shinn P, Jadhav A, Muller R, Waters KM, Shi W, Liu X, Zhang L, Knight R, Jensen PR, Palsson BO, Pogliano K, Linington RG, Gutierrez M, Lopes NP, Gerwick WH, Moore BS, Dorrestein PC, Bandeira N (2016) Sharing and community curation of mass spectrometry data with Global Natural Products Social Molecular Networking. Nat Biotech 34(8):828–837. https://doi.org/10.1038/nbt.3597 http://www.nature.com/nbt/journal/v34/n8/abs/nbt.3597.html—supplementary-information

Chapter 13

Infant Mouse Model of *Vibrio cholerae* Infection and Colonization

Jyl S. Matson

Abstract

Cholera is an epidemic diarrheal disease caused by *Vibrio cholerae* that continues to cause significant morbidity and mortality in many parts of the world. Several different animal models have been used by scientists over the years to study the pathogenesis of cholera. However, the most commonly used is the infant (suckling) mouse model, which has been found to replicate important aspects of human intestinal colonization. Here we present a detailed protocol for using the infant mouse model to assess the colonization of *V. cholerae* strains using a competition assay.

Key words *Vibrio cholerae*, Mouse model, Competitive index, Colonization

1 Introduction

Vibrio cholerae, the causative agent of epidemic cholera, colonizes the human small intestine where it produces the virulence factors responsible for the disease. Several different animal models have been used to attempt to replicate the human disease, with the aim of better understanding the mechanisms of pathogenesis and to identify virulence factors produced by the bacteria [1]. The most commonly used model is the infant (suckling) mouse model [2]. While adult mice are not able to be colonized by *V. cholerae* without elimination of the microbiota [3], infant mice (3–5 days old) are efficiently colonized. The reason for colonization of infant mice is not well understood, but thought to be due to the relative immaturity of the immune response [1].

The infant mouse model has been used to identify several important virulence factors in *V. cholerae*. Early studies showed that spontaneous rough strains of *V. cholerae* are defective in colonization of the infant mouse [4]. Later, strains with mutations in the LPS O-antigen were constructed and found to also have colonization defects [5, 6]. The infant mouse model was also used to identify other virulence factors including the accessory colonization

Aleksandra E. Sikora (ed.), *Vibrio Cholerae: Methods and Protocols*, Methods in Molecular Biology, vol. 1839,
https://doi.org/10.1007/978-1-4939-8685-9_13,

factors [7], a hemagglutinin [8], and several metabolic proteins [4, 5, 9, 10]. Most notably, the toxin-coregulated pilus (TCP) was originally identified as a colonization factor using this model [11]. Subsequently, TCP and its regulator ToxR were found to be essential colonization factors in a study using human volunteers, lending validity to the use of the infant mouse as a model for human colonization studies [12].

The most common application of the infant mouse model is that of a competition assay. These "competitive index" experiments, in which the colonization of a mutant *V. cholerae* strain is compared to that of a wild-type strain in the same mouse, have been used for decades to identify factors that contribute to the virulence of *V. cholerae* (also described in [13]). The mutant strain is generally *lacZ*$^+$, whereas the wild-type strain is *lacZ*$^-$. This permits determination of the number of bacteria from each strain present in the intestine of an individual mouse by plating the colonizing bacteria from the mouse intestine on selective media containing X-gal. These experiments are internally controlled, with the wild-type and mutant bacteria being delivered in the same number to the same animal. However, it is often beneficial to include a small number of control animals that are given a mixture of wild-type bacteria and a mutant that is known to be defective in colonization (such as a *toxT* mutant [14]).

While the infant mouse model can be used to assess the cholera pathogenesis in a number of ways, including lethality, in vivo expression technology (IVET), or signature-tagged mutagenesis (STM) experiments [5, 9, 11, 15], this chapter will focus on the method for the commonly used competition assay.

2 Materials

Prepare all solutions and reactions using ultrapure water and analytical grade reagents. Follow all appropriate waste disposal regulations when disposing waste materials.

1. *Vibrio cholerae* strains to be tested, streptomycin resistant (*see* **Note 1**).
2. LB media: 10 g tryptone, 5 g yeast extract, 5 g NaCl in 1 L water. Autoclave to sterilize.
3. LB plates: LB media containing 15 g agar per L. Autoclave to sterilize, and add streptomycin and X-gal before pouring plates, as appropriate.
4. 1000× streptomycin (100 mg/mL in water). Filter sterilize.
5. 1000× X-gal (5-Bromo-4-chloro-3-indolyl β-D-galactopyranoside; 40 mg/mL in DMF).
6. 1 mL tuberculin syringes.
7. 29 gauge needles.

8. Polyethylene tubing (PE 10), such as Becton Dickinson #427401.
9. Evan's Blue Dye (1% in water). Filter sterilize.
10. Isofluorane.
11. 30 °C incubator.
12. Dissection tools.
13. Tissue homogenizer(s) (*see* **Note 2**).
14. CD-1 infant mice (3–5 days old) (*see* **Note 3**).

3 Methods

Carry out all procedures at room temperature unless otherwise specified. All animals must be maintained in appropriate BSL-2 facilities during infection.

3.1 Preparation of Inoculum and In Vitro Control Cultures

1. Prepare overnight cultures of *V. cholerae* strains to be completed by inoculating one colony of each strain into LB media containing 100 μg/mL streptomycin. Incubate overnight with shaking at 37 °C.
2. Dilute each strain 1:50 in LB, and determine the OD_{600} of each strain the next day using a spectrophotometer. If the OD_{600} of the strains are not equal, dilute the more concentrated strain in LB until the OD_{600} matches the less concentrated strain.
3. Mix together 1 mL of each of the two strains to be competed and add 10 μL Evan's blue dye (*see* **Note 4**).
4. Prepare in vitro control culture by diluting 50 μL of the prepared inoculum mixture into 5 mL LB containing streptomycin. Incubate overnight (or for a time frame equal to the mouse infection period) with shaking at 37 °C.
5. Prepare serial dilutions of the cultures as described below for the intestinal homogenates and plate on LB agar containing streptomycin and X-gal for enumeration. After overnight incubation, count the number of blue and white colonies. Calculate the ratio of the mutant to wild-type strain after in vitro competition.

3.2 Infection of Infant Mice by Oral Gavage

1. Cut a length of the polyethylene tubing (10 cm or greater). Thread one end onto the needle attached to the syringe, taking care to not puncture the tubing. Draw up prepared inoculum mixture to load the syringe. Prepare one syringe for each competition.
2. Prepare a chamber for isoflurane sedation. An empty pipette tip box containing Kimwipes to absorb the isoflurane works well for this purpose (*see* **Note 5**).

3. Add a small amount of isoflurane (1 mL or less) to the Kimwipes in the bottom of the chamber. Lightly anesthetize mice singly or in small batches just until they stop moving.
4. Orally gavage mice by carefully inserting the tubing into the mouth and applying gentle pressure to guide the tubing into the throat (*see* **Note 6**). Press on the syringe to deliver 50 μL of the inoculum directly into the stomach.
5. After inoculation, pups that receive the same treatment should be grouped together in 250 mL plastic beakers containing shredded paper towels or appropriate disposable containers. Place mice in a 30 °C incubator, and monitor closely for the first 2 h after inoculation to ensure that the mice have recovered from the procedure (*see* **Note 7**).

3.3 Harvesting and Plating of Bacteria Colonizing the Mouse Intestinal Tract

1. Euthanize the pups by cervical dislocation (or other approved method, *see* **Note 8**) on the day following infection (15–24 h post-gavage).
2. Dissect each mouse by cutting the abdomen longitudinally with a scissors or scalpel from beneath the ribcage to the bottom of the abdomen. Care should be taken to not cut into the internal organs.
3. Remove the entire intestinal tract from below the stomach to the anus from each mouse using a forceps, and place in 5 mL LB on ice.
4. Homogenize the intestine to break up visible tissue.
5. Prepare serial dilutions of each homogenized sample. Dilute into LB and maintain on ice until plating. Four tenfold dilutions of each sample are typically sufficient to obtain countable numbers of colonies per plate.
6. Spread 100 μL of each dilution on LB agar plates containing streptomycin (to select for *V. cholerae*) and X-gal (to distinguish wild type from mutant bacteria). Incubate plates overnight at 37 °C.
7. Count the numbers of blue and white colonies the next day on those plates with a total number of colonies between 30 and 300. If more than one plate per mouse falls within this range, average the numbers from the countable plates.
8. Calculate the competitive index for each mutant strain tested by dividing the number of recovered blue colonies by the number of white (wild type). If the ratio of the mutant is significantly different than that of the wild-type strain recovered from the in vitro competition, the overall competitive index should be adjusted accordingly, or both in vitro and in vivo competitive indexes should be provided for comparison.

4 Notes

1. Using streptomycin-resistant strains allows for the selection of *V. cholerae* over the background of the mouse intestinal microbiota.
2. Several types of tissue homogenizers/grinders can be used at this step as long as they adequately disrupt the intestinal tissue. If a handheld electric homogenizer is used, it is important to wash the probe in ethanol and water between samples to avoid cross-contamination. Disposable plastic or reusable glass mortar-pestle-type grinder/homogenizers can also be used as long as they are large enough to process a 5 mL sample.
3. We typically obtain the mice from an approved supplier that offers litters of CD-1 infants for direct purchase, along with the dams. This limits the need for housing, husbandry, and veterinary care for the animals and enables considerable savings on the costs of these experiments. We typically receive approximately ten pups with each dam, but there can be some variation in the number of pups per litter. Alternatively, pregnant females can be purchased, and the animals can be housed and maintained until after the pups are delivered.
4. We include Evan's blue dye in the inoculum to verify that the inoculum is delivered properly to the stomach (this makes the inoculum in the stomach very obvious as the mice are still hairless). Any mice where the inoculum has been improperly delivered into the lungs should be immediately euthanized.
5. Isoflurane works well in infant mice and prevents them from struggling during the gavage, reducing chances of injury or improper delivery of the inoculum. However, it is important to only lightly sedate the animals to prevent overdose.
6. We find it helpful to place a mark on the tubing approximately 1 cm from its end to use as a reference point during the gavage. Once the tubing has been inserted into the throat to roughly that point, the inoculum should be delivered.
7. The pups must be separated from the dams during this incubation period, as feeding after inoculation interferes with *V. cholerae* colonization and causes the results to be non-reproducible. The infant mouse model for *V. cholerae* has been used for over 30 years, and it has been experimentally determined that isolation of the pups from the dams is crucial for successful colonization. The pups require incubation at 30 °C overnight to maintain their body temperatures while separated from the dams.
8. Asphyxiation by CO_2 overdose is not recommended as a euthanasia procedure for infant mice as they are relatively resistant to hypoxia.

References

1. Richardson SH (1994) Animal models in cholera research. In: Wachsmuth IK, Blake PA, Olsvik Ø (eds) *Vibrio cholerae* and cholera: molecular to global perspectives. ASM Press, Washington, DC
2. Klose KE (2000) The suckling mouse model of cholera. Trends Microbiol 8:189–191
3. Butterton JR, Ryan ET, Shahin RA, Calderwood SB (1996) Development of a germfree mouse model of *Vibrio cholerae* infection. Infect Immun 64:4373–4377
4. Baselski VS, Upchurch S, Parker CD (1978) Isolation and phenotypic characterization of virulence-deficient mutants of *Vibrio cholerae*. Infect Immun 22:181–188
5. Chiang SL, Mekalanos JJ (1998) Use of signature-tagged transposon mutagenesis to identify *Vibrio cholerae* genes critical for colonization. Mol Microbiol 27:797–805
6. Iredell JR, Stroeher UH, Ward HM, Manning PA (1998) Lipopolysaccharide O-antigen expression and the effect of its absence on virulence in *rfb* mutants of *Vibrio cholerae* O1. FEMS Immunol Med Microbiol 20:45–54
7. Hughes KJ, Everiss KD, Kovach ME, Peterson KM (1995) Isolation and characterization of the *Vibrio cholerae acfA* gene, required for efficient intestinal colonization. Gene 156:59–61
8. Franzon VL, Barker A, Manning PA (1993) Nucleotide sequence encoding the mannose-fucose-resistant hemagglutinin of *Vibrio cholerae* O1 and construction of a mutant. Infect Immun 61:3032–3037
9. Camilli A, Mekalanos JJ (1995) Use of recombinase gene fusions to identify *Vibrio cholerae* genes induced during infection. Mol Microbiol 18:671–683. https://doi.org/10.1111/j.1365-2958.1995.mmi_18040671.x
10. Henderson DP, Payne SM (1994) *Vibrio cholerae* iron transport systems: roles of heme and siderophore iron transport in virulence and identification of a gene associated with multiple iron transport systems. Infect Immun 62:5120–5125
11. Taylor RK, Miller VL, Furlong DB, Mekalanos JJ (1987) Use of *phoA* gene fusions to identify a pilus colonization factor coordinately regulated with cholera toxin. Proc Natl Acad Sci U S A 84:2833–2837
12. Herrington DA, Hall RH, Losonsky G, Mekalanos JJ, Taylor RK, Levine MM (1988) Toxin, toxin-coregulated pili, and the *toxR* regulon are essential for *Vibrio cholerae* pathogenesis in humans. J Exp Med 168:1487–1492
13. Angelichio MJ, Spector J, Waldor MK, Camilli A (1999) *Vibrio cholerae* intestinal population dynamics in the suckling mouse model of infection. Infect Immun 67:3733–3739
14. Champion GA, Neely MN, Brennan MA, DiRita VJ (1997) A branch in the ToxR regulatory cascade of *Vibrio cholerae* revealed by characterization of *toxT* mutant strains. Mol Microbiol 23:323–331
15. Lee SH, Hava DL, Waldor MK, Camilli A (1999) Regulation and temporal expression patterns of *Vibrio cholerae* virulence genes during infection. Cell 99:625–634

Chapter 14

Methods for Assessments of Collagenolytic Activity of the *Vibrio cholerae* Extracellular Proteases, Purification of Secreted Collagenase VchC, and Extraction of Type I Collagen from Fish Skin

Bo R. Weber and Aleksandra E. Sikora

Abstract

Secreted proteases have been linked to facilitating the survival of *Vibrio cholerae* in different environmental niches. Examination of protease activity using various measures is critical to determine the substrate specificity of a given proteolytic enzyme and its biological function. Here, we describe methods to assess collagenolytic activity of the *V. cholerae* extracellular proteases using synthetic and natural substrates in quantitative and qualitative assays. Purification of the secreted *V. cholerae* collagenase VchC and extraction of type I collagen from fish skin are also described.

Key words Metalloprotease, Collagenolytic proteases, Secretion, Collagenase, Type I collagen, SDS-PAGE, FALGPA, Zymography, Affinity chromatography, DQ-gelatin

1 Introduction

Collagen, or matrix protein, is one of the most abundant extracellular structural proteins in metazoans and provides a rich nutrient source for bacteria that produce collagenolytic enzymes [1]. Due to its complicated hierarchical structure and insolubility in water, only a few proteases with unique features possess the ability to degrade collagens. The true collagenases are enzymes that directly hydrolyze native collagen molecules with unique specificity and should be differentiated from other collagen-degrading proteases and from gelatinases, which decompose gelatin, a denatured form of collagen [1–3]. The relationship between collagen and *V. cholerae* has recently been discovered by identification and characterization of a secreted metalloprotease VchC displaying collagenolytic activity [4]. Other metalloproteases with gelatinase activity include HapA and PrtV, which degrade chironomid egg masses and kill grazing predators such as worms and protozoans [5, 6], respectively.

Aleksandra E. Sikora (ed.), *Vibrio Cholerae: Methods and Protocols*, Methods in Molecular Biology, vol. 1839,
https://doi.org/10.1007/978-1-4939-8685-9_14,

To study the association between collagen and the metalloproteases secreted by *V. cholerae*, both quantitative and qualitative methods can be applied. This chapter focuses on assays developed to determine protease specificity using a synthetic peptide specific for collagenases, *N*-(3-(2-furyl)acryloyl)-Leu-Gly-Pro-Ala (FALGPA), and natural substrates such as type I fish and human collagens. Purification of type I collagen from fish skin and assessment of its purity are also delineated. In addition, we offer detailed description of purification of extracellular collagenase VchC from *V. cholerae* culture supernatant.

1.1 Gelatin Zymography

Gelatin zymography is a technique utilized to detect the proteolytic activity specific to metalloproteases, collagenases, or other proteases against gelatin in the separating gel. This method allows an assessment of the gelatinolytic activity of purified proteases, enzymes in culture supernatants or in other samples. The advantage of gelatin zymography, in contrast to the enzymatic assay with DQ-gelatin or FALGPA, is an examination of molecular weight and the visualization of variant forms of the hydrolases present in a sample [7]. Samples of a purified protease or supernatants isolated from *V. cholerae* (e.g., at distinct time points of bacterial growth or different strains) are matched by the protein concentration or equivalent bacterial density (optical density at 600 nm; OD_{600}), respectively, separated on zymogram gels, and the proteins in a gel are renaturated by incubation in a buffer supplemented with calcium chloride and glycerol. After the refolding of proteins, the zones of gelatinolytic activity are revealed by staining the gel with colloidal Coomassie. The activity of a protease(s) is observed as a clearing band, where the gelatin was digested by the protease against the dark background (Fig. 1).

1.2 Assessment of Proteolytic Activity Using a Highly Quenched, Fluorescein-Labeled Gelatin from Pig Skin: DQ-Gelatin

The enzymatic activity of a purified protease or extracellular enzymes in a culture supernatant can be tested by using a highly quenched, fluorescein-labeled gelatin from pig skin (DQ-gelatin) as a substrate. This method enables the detection of both collagenase and gelatinase activity with high sensitivity. The proteolytic activity is measured as the linear change in fluorescence (ΔFU) over time at 495 nm excitation and 535 emission wavelengths using 96-well black nonbinding surface microtiter plates and a fluorescence microplate reader and normalized by either the protein amount or the density of the culture (Fig. 2). The controls for the assay may include collagenase purified from Collagenase D from *Clostridium histolyticum* and a general inhibitor of metalloproteases, 1,10-phenanthroline.

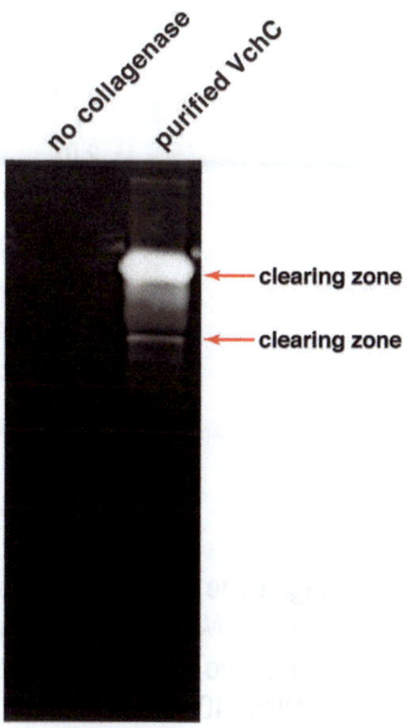

Fig. 1 Example of gelatin zymography. Enzymatic activity of VchC was examined by *in gel* zymography. The purified VchC was separated in 10% Tris-glycine gel copolymerized with 0.1% gelatin. Red arrows labeled pointing to clearing zone indicate two detectable forms of VchC

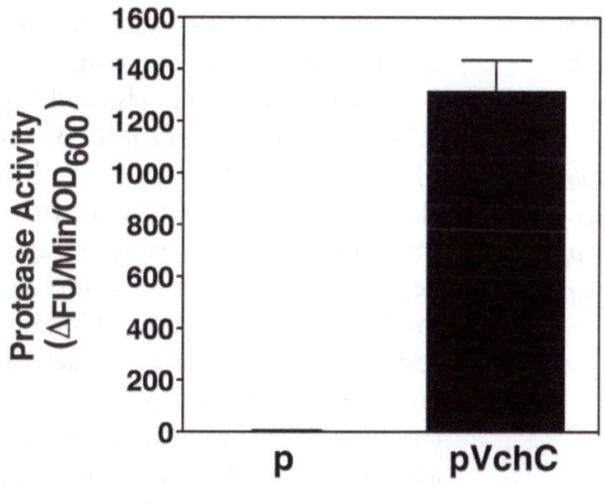

Fig. 2 Example of measuring proteolytic activity in *V. cholerae* culture supernatants using DQ-gelatin as a substrate. The supernatants of *V. cholerae* N16961 wild type bearing either pMMB67-EH (p) or plasmid-borne VchC (pVchC) were harvested from cultures grown at 37 °C in LB medium supplemented with IPTG to stationary phase of growth. The detected collagenolytic/gelatinolytic activity is expressed as a change in fluorescence (ΔFU) over time (Min.) and normalized by optical density (OD_{600}) of corresponding bacterial cultures. Experiments were performed on three separate occasions in technical triplicates, and the mean with corresponding standard errors of the mean (SEM) is presented

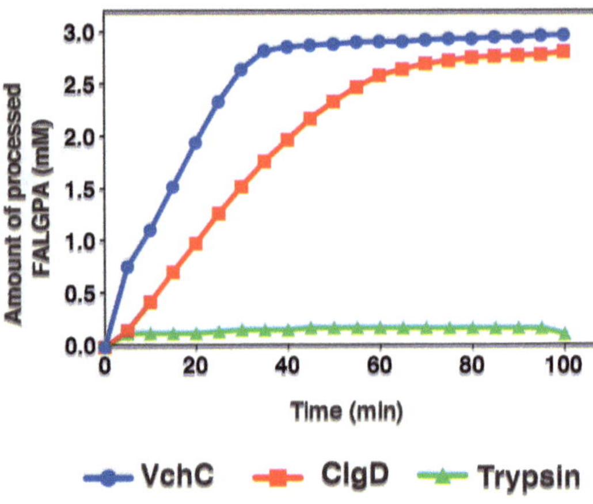

Fig. 3 The ability of protease to cleave a collagenase specific substrate, synthetic peptide *N*-(3-(2-furyl)acryloyl)-Leu-Gly-Pro-Ala (FALGPA) is presented based on VchC. The purified VchC at 60 μg/mL was added to 4.0 mM FALGPA in 50 mM Tricine, 400 mM NaCl, 10 mM $CaCl_2$, pH 7.5. The reactions were incubated at 37 °C for 5 min, followed by monitoring the change in absorbance at 345 nm for 1 h every 5 min at 30 °C. The amount of degraded FALGPA was calculated based on the calibration curve. Purified trypsin (green) and Collagenase D from *C. histolyticum* (red; ClgD) served as a negative and a positive control for the assay, respectively

1.3 Assessment of Collagenolytic Activity Using a Synthetic Collagenase Substrate, 2-Furanacryloyl-L-Leucylglycyl-L-Prolyl-L-Alanine (FALGPA)

The collagenolytic activity of protease can be assessed in a quantitative assay using a synthetic peptide substrate, 2-furanacryloyl-L-leucylglycyl-L-prolyl-L-alanine (FALGPA) [8]. FALGPA is cleaved rapidly by collagenase according to Michaelis-Menten kinetics, whereas other proteases such as trypsin, chymotrypsin, thermolysin, or elastase have no ability to degrade this substrate. The assay is rapid, convenient, and sensitive. The negative and positive controls for the assays should include modified trypsin and Collagenase D from *C. histolyticum*, respectively. The reactions are initiated by the addition of purified enzymes (the tested purified protease, trypsin, or Collagenase D) to samples containing FALGPA. Subsequently, the decrease in absorbance of the furanacryloyl peptide associated with hydrolysis of the peptide bond between the first and the second residue is monitored at 345 nm using a microplate reader. The amount of degraded FALGPA should be calculated based on the calibration curve obtained by measuring the absorbance of the reference samples, comprised of increasing concentrations of FALGPA, in the reaction buffer. The collagenolytic activity is evident when compared to the positive and the negative controls (Fig. 3).

1.4 Purification of a Secreted, Chimeric VchC from V. cholerae Culture Supernatants

We describe an efficient strategy for obtaining a highly pure *V. cholerae* collagenase VchC, which can be adopted to other proteases. The gene-encoding VchC (containing the native signal peptide, ribosome-binding site, and the C-terminal 6× His-epitope) was cloned under the control of an isopropyl-beta-D-thiogalactopyranoside (IPTG)-inducible promoter placed on a broad-host vector, pMMB67-EH. This strategy enables overproduction of VchC and its subsequent extracellular translocation via type II secretion pathway in *V. cholerae* N16961 [4]. The culture supernatant containing the chimeric protein is harvested when the bacteria reach an early stationary stage of growth. The proteins in the supernatant are precipitated by saturation with ammonium sulfate [9], and the protease, after dialysis, is purified to homogeneity using affinity chromatography.

1.5 Purification of Type I Collagen from Fish Skin and Preparation of Human Type I Collagen

Type I collagen belongs to the group of fibrillar collagens that are spread from sponges to humans and constitute over 95% of total collagens in many animal tissues [10, 11]. Type I collagen can be extracted in a highly pure form by subjecting fish skin (e.g., from trout) to removal of non-collagenous proteins with sodium peroxide and by defatting fish skin with butyl alcohol, acetic acid treatment, precipitation, and dialysis [4, 12]. The purity of extracted collagen can be verified by in sodium dodecyl sulfate polyacrylamide protein electrophoresis (SDS-PAGE) and Coomassie staining (Fig. 4). Human collagen naturally secreted from human neonatal fibroblast cells that comprise type I collagen (97%) and type III collagen (3%) is commercially available, but the solvent (0.1 N HCl) precludes its utilization in downstream enzymatic assays. Therefore, we also describe the dialysis of this reagent against a buffer, allowing assessment of collagenase hydrolytic activity.

1.6 Assessment of Collagenolytic Activity Using Type I Collagens

Type I collagen from fish skin and secreted from human neonatal fibroblast cells (prepared as described above) can be examined as the candidate substrate proteins of collagenase. In these assays, a certain amount of a purified protease (e.g., 2 pmol) or culture supernatants are incubated with collagens followed by evaluation of the digestion by SDS-PAGE coupled with Coomassie staining (Fig. 4).

1.7 SDS-PAGE and Coomassie Staining

A downstream step used for analysis of collagenase purity and enzymatic activity against type I collagens or other natural substrates is SDS-PAGE that separates proteins according to molecular weight and is coupled with protein visualization using Coomassie staining. This is the method of choice because it enables verification of all samples in a rapid, cost-effective, and straightforward manner.

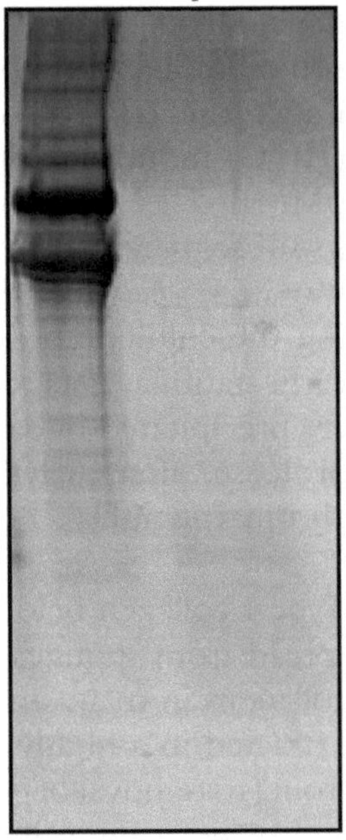

Fig. 4 Examination of purity of purified type I fish collagen and collagenase activity using SDS-PAGE and Coomassie staining. Purified type I collagens from fish skin were incubated with (+) and without purified VchC (2 pmol) for 1 h at 37 °C. The samples were subsequently resolved by 10% SDS-PAGE, and proteins were stained with colloidal Coomassie

2 Materials

Disposable gloves, tips, pipettes, and 1.5 mL centrifuge tubes (e.g., Eppendorf tubes) are needed for all procedures.

2.1 Subfractionation of Culture Supernatant

1. Growth of *V. cholera*.
2. LB agar (e.g., Difco).
3. LB medium (e.g., Difco).
4. Antibiotics (e.g., Teknova, Amresco).
5. Petri dishes.
6. 50 mL conical tubes.

2.2 Purification of Native Protease from *V. cholerae*

1. Buffer A: 20 mM Tris–HCl pH 8.0, 150 mM NaCl, 1 mM imidazole.
2. Buffer B: 20 mM Tris–HCl pH 8.0, 300 mM Nacl, 10 mM imidazole.
3. Buffer C: 20 mM Tris–HCl pH 8.0, 500 mM NaCl, 250 mM imidazole.

4. Buffer D: 20 mM Tris–HCl pH 8.0, 1 mM $CaCl_2$, 10% glycerol.
5. Nickel-nitrilotriacetic acid (Ni-NTA column): 500 μL of Ni-NTA resin, lysis buffer for equilibration of the column.
6. Snakeskin dialysis tubing (ThermoFisher).
7. Ammonium sulfate (*see* **Notes 1** and **6**): 80.5 g ammonium sulfate per 250 mL of supernatant (placed in 500 mL bottles). Stir at 4 °C until ammonium sulfate is fully solubilized, then leave to stir for 1 h.

2.3 Measuring Concentration of Purified Native Protease (See Note 2)

For example, BioRad DC Protein Assay kit.

2.4 SDS-PAGE

1. Loading buffer: LDS loading buffer (e.g., Invitrogen, BioRad), 50 mM dithiothreitol.
2. Hand-casted 10% SDS-PAGE (or precast gels): Resolving gel (4.85 mL ddH_2O, 2.5 mL 40% acrylamide-bis, 100 μL 10% SDS, 2.5 mL Tris 1.5 M pH 8.8, 50 μL 10% APS (mg/μL), 5 μL TEMED). Stacking gel (3.18 mL ddH_2O, 0.5 mL 40% acrylamide-bis, 50 μL 10% SDS, 1.26 mL Tris 0.5 M pH 6.8, 25 μL 10% APS, 5 μL TEMED).
3. 10% SDS-PAGE (e.g., Invitrogen): Novex Tris-glycine SDS (10×), Novex Tris-glycine SDS sample buffer (2×).
4. Colloidal Coomassie: 35.5 mL 85% phosphoric acid (H_3PO_4), 170 g ammonium sulfate ($(NH_4)_2SO_4$), 340 mL methanol, 1 g Coomassie Brilliant Blue G-250.

2.5 Zymography

1. 0.1% gelatin in 10% Tris-glycine gel (e.g., Novex, Invitrogen): Tris-glycine SDS sample buffer (2×), Tris-glycine SDS running buffer (10×), Novex Zymogram renaturing buffer (10×), and Novex zymogram developing buffer (10×).
2. Colloidal Coomassie.

2.6 Fluorescein Conjugate Gelatin

1. DQ-gelatin from pig skin (e.g., ThermoFisher) (*see* **Note 3**).
2. DQ-gelatin buffer (50 mM Tris–HCl pH 7.6, 150 mM NaCl, 5 mM $CaCl_2$).
3. Aluminum foil.

2.7 Purified Type I Fish Collagen

1. One rainbow trout (*see* **Note 4**).
2. 0.1 N NaOH.
3. 10% butyl alcohol.
4. 0.5 M acetic acid.

5. Buffer E (2.6 M NaCl, 0.05 M Tris–HCl pH 7.5).
6. 0.1 M acetic acid.
7. Snakeskin dialysis tubing (ThermoFisher).

2.8 Measuring Concentration of Purified Type I Fish Collagen

1. BioRad DC Protein Assay kit.

2.9 Preparation of Type I Human Collagen

1. Human collagen, naturally secreted from human neonatal fibroblast cells (e.g., VitroCol, Advanced BioMatrix) comprising of type I collagen (97%) and type III collagen (3%).
2. 0.1 N HCl.
3. Buffer F (20 mM Tris–HCl pH 8.0, 1 mM $CaCl_2$).
4. Snakeskin dialysis tubing (ThermoFisher).

2.10 Substrate Specificity Assay Using Collagenase-Specific Substrate

1. *N*-(3-(2-furyl)acryloyl)-Leu-Gly-Pro-Ala (FALGPA; e.g., Bachem AG).
2. Modified trypsin (e.g., New England Biolabs).
3. Collagenase D from *C. histolyticum* (e.g., Roche, ThermoFisher).
4. Purified native protease, e.g., VchC.
5. Buffer G: 50 mM Tricine pH 7.5, 400 mM NaCl, 10 mM $CaCl_2$.

2.11 Substrate Specificity Assay Using Type I Human and Fish Collagen

1. Purified type I collagen from fish skin.
2. Purified type I human collagen from fibroblasts.
3. Purified native protease.
4. Butter C.

2.12 Equipment

1. Gel buffer tank (e.g., BioRad, Invitrogen).
2. Gel staining box.
3. Power supply.
4. 1.5 mL centrifuge tubes (e.g., Eppendorf tubes).
5. 50 mL conical tubes.
6. 250 mL centrifuge bottles.
7. 500 mL centrifuge bottles.
8. Tube closures.
9. Thermoblock.
10. Gel documentation system (e.g., BioRad ChemiDoc).
11. Plate reader (for measuring absorbance and fluorescence; e.g., Synergy HT; BioTek).

12. Incubator.
13. A bench-top or floor model centrifuge with a rotor capable of accommodating 500 or 250 mL centrifuge bottles as well as 50 mL conical tubes (e.g., Beckman Coulter Avanti J-E).
14. Microfuge.
15. Pipettors.
16. Centrifugation rotors (e.g., JS-5.3 and JLA-10.500; Beckman Coulter).

3 Methods

3.1 Preparation of Culture Supernatants [4]

Grow all cultures at 37 °C with aeration.

1. Streak required *V. cholerae* strain on LB agar containing appropriate antibiotic and incubate for 16 h.
2. Using a single colony, inoculate 5 mL of LB medium supplemented with appropriate antibiotic in a 50 mL conical tube.
3. Culture the bacteria with aeration for 16 h.
4. Measure optical density at 600 nm wavelength (OD_{600}).
5. Centrifuge the culture at 4000 × *g* for 10 min.
6. Carefully collect culture supernatant into 15 mL tube and place it on ice.
7. Proceed with downstream application, e.g., measurement of protease activity.
8. Store supernatants in 1.5 mL aliquots at −20 °C (*see* **Note 5**).

3.2 Purification of the *V. cholerae* Extracellular Collagenase VchC from Culture Supernatants [4]

Perform all of the following procedures at 4 °C, unless otherwise specified.

Preparation of the Ni-NTA column for gravity purification.

1. Assemble a 10 mL Ni-NTA column (e.g., ThermoScientific).
2. Place 0.5 mL of Ni-NTA resin in a 15 mL conical tube and wash five times with 5 mL of buffer A.
3. Add 6 mL of buffer A; gently suspend resin and apply to column.
4. Allow resin to sediment (e.g., overnight).
5. Wash the column with six bed volumes of buffer A prior to loading a sample.

Purification of protease from *V. cholerae*.

1. Streak frozen stock of *V. cholerae* N16961 harboring pVchC-His (or another construct expressing the desired protease) on LB agar supplemented with an appropriate antibiotic(s).

2. Incubate plates for 16 h at 37 °C.
3. Inoculate with a single colony 5 mL LB medium supplemented with appropriate antibiotic and culture for 16 h at 37 °C with aeration.
4. Dilute the overnight culture 1:100 into 1 L of LB medium supplemented with appropriate antibiotic and 100 μM IPTG.
5. Culture bacteria with aeration at 37 °C for ~4 h until the early stationary phase of growth is reached (OD_{600} of ~4).
6. Centrifuge cultures in 250 mL centrifugation bottles at 4000 × *g* for 20 min.
7. Collect the culture supernatant and passage through 0.22 μM filters (e.g., VWR).
8. Divide supernatant into ~250 mL aliquots in four 500 mL centrifugation bottles.
9. Precipitate proteins at 40% saturation of ammonium sulfate by adding 80.5 g of ammonium sulfate into each bottle in small quantities and dissolve using a stirring bar (*see* **Notes 1** and **6**).
10. Once the ammonium sulfate has completely dissolved, allow to stir for 1 h.
11. Remove stirring bars.
12. Centrifuge culture supernatants at 10,000 × *g* for 40 min.
13. Decant supernatant and resuspend the pellet in each bottle in 2–4 mL of buffer A (*see* **Note 6**).
14. Transfer suspensions to the dialysis tubing using a 1000 μL pipette tip (*see* **Note** 7).
15. Dialyze in 2 L of buffer A three times for 30 min each (*see* **Note 8**).
16. Apply dialyzed sample to prepared Ni-NTA column.
17. Collect the flow-through into a 50 mL conical tube.
18. Wash resin with 5 mL of buffer B, and collect flow-through in a 15 mL conical tube.
19. Repeat washes four times with the same amount of buffer B.
20. Elute bound proteins with 3 mL of buffer C collecting six 500 μL fractions into 1.5 mL centrifugation tubes (e.g., Eppendorf tubes).
21. Analyze the purity of purified protease by SDS-PAGE.
22. Combine the elution fractions containing the purified protein, and concentrate with exchange buffer C using a centrifugal 5 mL tube with appropriate molecular mass cutoff (e.g., Microsep Advance Pall Life Sciences).
23. Exchange buffer five to six times with a new portion of buffer C by centrifuging the sample at 5000 × *g* for 10–15 min until NaCl and imidazole concentrations are minimal.

24. Exchange buffer C with buffer D by centrifuging the sample at 5000 × *g* for 10–15 min and store the protein at −80 °C.

3.3 Measuring Protein Concentration (See Note 2)

Follow instructions below, adapted from the manufacturer's recommendation (BioRad DC Protein Assay).

1. Prepare 10 μL volumes of undiluted purified protein and the purified protein diluted to the following: twofold, fivefold, and tenfold.
2. Using a clean dry clear flat-bottom immune nonsterile 96-well microtiter plate (e.g., ThermoScientific), add 5 μL of ddH_2O as a negative control and increasing concentrations of BSA (0.2–0.9 mg/mL included in BioRad DC Protein Assay kit).
3. Add 5 μL of samples prepared in **step 1**.
4. Add 25 μL Reagent A into each sample.
5. Add 200 μL Reagent B into each sample.
6. Allow samples to mix gently on rotor and incubate at 25 °C for 15 min.
7. Measure absorbance at 750 nm using a microplate reader (e.g., Synergy HT Multi-Mode Microplate Reader; BioTek).

3.4 SDS-PAGE

For all analyses, normalize the samples by OD_{600} of bacterial cultures or protein concentration.

Preparation of hand-casted 10% SDS-PAGE.

All steps should be completed at room temperature.

1. Following manufacturer's instructions, assemble the hand-casting chamber (BioRad).
2. Mix reagents for resolving gel in the order of reagents provided in Subheading 2.4 in a 50 mL glass beaker (*see* **Note 10**).
3. Pour resolving gel from the side lip of the beaker until the mixture has reached the bottom of the green line of the casing.
4. Add 1 mL ddH_2O above the resolving gel to prevent gel dehydration. Allow to polymerize for 45 min.
5. Mix reagents for stacking gel in the order of reagents provided in Subheading 2.4 in a 50 mL glass beaker.
6. Decant ddH_2O from above the resolving gel.
7. Pour stacking gel from the side lip of the beaker and place the desired 10- or 15-well comb into the casting.
8. Allow to polymerize for 45 min and use immediately or store at 4 °C in a tight casting that maintains gel hydration (*see* **Note 11**).

3.5 Running SDS-PAGE and Preparing Samples

10% SDS-PAGE or precast gel (e.g., BioRad, ThermoFisher).

Follow manufacturer's recommendations to assemble the buffer tank and gel cassette. We recommend BioRad and Invitrogen systems. The following procedure describes running SDS-PAGE using the XCell *SureLock* Mini-Cell (Invitrogen).

1. Assemble a precast gel into XCell *SureLock* Mini-Cell (Invitrogen).
2. Use 100 mL 10× Novex Tris-glycine SDS and 900 mL of ddH_2O to dilute to 1× (*see* **Note 12**).
3. Mix the buffer and fill buffer tank according to the manufacturer's instruction.
4. Aliquot the desired volume of each sample normalized based on OD_{600} or protein concentration into 1.5 mL centrifuge tubes.
5. Add 5 μL of 2× Novex Tris-glycine SDS sample buffer and appropriate amount of ddH_2O to reach a total volume of 10 μL.
6. Heat the samples at 85 °C for 5 min.
7. Load the appropriate concentration of your protein sample on the gel immediately.
8. Run the gel at 125 V constant for 45 min.

3.6 Zymography

The proteolytic activity of a protease against 0.1% gelatin as a substrate in 10% Tris-glycine gel (e.g., Novex, Invitrogen) can be assessed. All steps should be performed at room temperature, unless otherwise specified.

1. Assemble zymogram gel in SDS-PAGE gel running system e.g., XCell *SureLock* Mini-Cell (Invitrogen).
2. Use 100 mL of 10× Tris-glycine SDS running buffer and add 900 mL ddH_2O to dilute buffer tenfold (*see* **Note 10**).
3. Fill gel box with 1× Tris-glycine SDS running buffer.
4. Prepare samples for zymography by adding the calculated amount of purified protein or supernatant, 5 μL of 2× Tris-glycine SDS sample buffer and appropriate amount of ddH_2O to reach a total volume of 10 μL.
5. Do not heat samples.
6. Load protein ladder and samples into gel.
7. Run gel at 125 V constant for 90 min.
8. Use 2.5 mL of 10× Novex zymogram renaturing buffer and 5 mL of 10× zymogram developing buffer, and add 22.5 mL and 45 mL ddH_2O to each buffer, respectively, to obtain 1× buffers for protein refolding (*see* **Note 13**).

9. After electrophoresis, carefully remove the gel from the cassette, and incubate the gel in 1× Zymogram renaturing buffer for 30 min with gentle agitation (*see* **Note 14**).
10. Remove the Zymogram renaturing buffer, and add 1× zymogram developing buffer.
11. Incubate the gel for 30 min with gentle agitation.
12. Remove the buffer and add fresh 1× zymogram developing buffer.
13. Incubate the gel at 37 °C for at least 4 h or overnight to refold the proteins in the gel (*see* **Note 15**).
14. Stain the gel in colloidal Coomassie Blue.

3.7 Fluorescein Conjugate Gelatin

The protease activity can be measured using DQ-gelatin from pig skin (e.g., ThermoFisher) in culture supernatants and using purified protease (e.g., 100 fmol) following the instructions provided by the manufacturer. The detected change in fluorescence should be normalized by either optical density of the cultures (OD_{600}) or protein concentration [13]. All steps should be performed at room temperature, unless otherwise specified.

1. Aliquot appropriate samples in total volumes of 100 μL and 50 μL for culture supernatant and purified protease (100 fmol), respectively, in an optical-bottom 96-well microtiter plate (e.g., ThermoScientific).
2. Include culture medium (e.g., LB) and buffer F as the background controls for supernatants and purified protease, respectively.
3. Incorporate appropriate vehicle controls if protease inhibitors solubilized in different solvents are included in the assays.
4. Prepare DQ-gelatin (EnzCheck, Molecular Probes) in 1 mL ddH_2O, and dilute using DQ-gelatin buffer fourfold and tenfold (*see* **Note 3**).
5. Utilize 1:4 ratio of fourfold and tenfold DQ-gelatin for each reaction (25 μg/mL DQ-gelatin).
6. Assess protease activity by recording the excitation and emission wavelengths at 485 ± 10 and 530 ± 15 nm, respectively, for 10 min at 37 °C using plate reader (e.g., the Synergy HT Multi-Mode Microplate Reader, BioTek).

3.8 Purification of Type I Fish Collagen [4]

After de-skinning perform all steps at 4 °C.

1. Purchase rainbow trout at local grocery store and remove fish skin using a razor blade.
2. Mix 0.1 N NaOH at sample to alkali solution ratio of 1:10 (w/v).

3. Incubate for 24 h, and exchange solution every 6 h.
4. Wash sample with cold ddH_2O until basic pH is reached.
5. Defat the sample with 10% butyl alcohol at sample to alkali solution ratio of 1:10 for 24 h and exchange solution every 6 h.
6. Repeat **step 4**.
7. Soak sample in 0.5 M acetic acid with solid to solvent ratio of 1:15 for 24 h.
8. Centrifuge mixture at 8000 × *g* for 30 min.
9. Save the supernatant.
10. Re-extract precipitate in 0.5 M acetic acid at sample to solution ratio of 1:30 (w/v) for 16 h with gentle stirring.
11. Repeat **step 8**.
12. Mix resulting supernatant with supernatant obtained in **step 9**.
13. Precipitate supernatant with buffer E via centrifugation at 8000 × *g* for 30 min.
14. Dissolve precipitate in ten volumes of 0.5 M acetic acid.
15. Transfer solution to dialysis tubing (*see* **Note 8**).
16. Dialyze sample against ten volumes of 0.1 M acetic acid for 12 h and change solution every 4 h.
17. Dialyze sample against ten volumes of ddH_2O until neutral pH is obtained (*see* **Note 16**).

3.9 Measuring Concentration of Purified Type I Collagen (See Note 17)

See Subheading 3.3 for measuring concentration of purified native protease.

3.10 Preparation of Type I Human Collagen

1. Solubilize human collagen in 0.1 N HCl.
2. Dialyze solubilized human collagen against buffer F using Amicon Ultra 0.5 mL filters.
3. Dilute the sample twofold in buffer F.

3.11 Substrate Specificity Assay Using Collagenase-Specific Substrate [4]

1. For culture supernatant, include growth medium (e.g., LB) as a control. For purified protease, include storage buffer (e.g., buffer F) as a control.
2. Dilute FALGPA to 2.0 mM using buffer G.
3. For each tested sample, include the reference samples composed of serially diluted FALGPA.
4. Incubate samples at 37 °C for 5 min in an optical-bottom 96-well microtiter plate (e.g., ThermoScientific).
5. Monitor the decrease in absorbance of the substrate at 345 nm every 5 min for 1 h in microplate reader set to 30 °C (e.g., the Synergy HT Multi-Mode Microplate Reader, BioTek).

6. The decrease in absorbance indicates protease activity (Fig. 3). The difference between the reference sample and the tested sample must be calculated to determine protease activity.

3.12 Substrate Specificity Assay Using Type I Human and Fish Collagen

All assays should be executed at 37 °C.

1. Dilute purified protease (e.g., 100 fmol) in buffer F in the following concentrations: 4, 10, 20, 40, and 80 fmol.
2. Add 5 μg of type I collagen of human or fish origin with increasing amounts of the purified protease (0, 4, 10, 20, 40, 80, and 100 fmol) (*see* **Note 17**).
3. Utilize SDS-PAGE and colloidal Coomassie staining to evaluate degradation of the tested protease substrates.

3.13 Preparation of Colloidal Coomassie Blue

1. Add 35.5 mL of 85% phosphoric acid to 420 mL of ddH_2O.
2. Add 170 g of ammonium sulfate and stir until completely dissolved.
3. Pour 340 mL of methanol and continue to stir while white precipitate forms (Solution A).
4. Make Solution B by adding 1 g of Coomassie Brilliant Blue G-250 into 100 mL of hot ddH_2O.
5. Heat Solution B until boiling.
6. Pour Solution B into Solution A, and add ddH_2O to 1 L.
7. Continue to stir for a couple of minutes. The stain will not dissolve well unless the volume and composition are accurate. Appropriately prepared Colloidal Coomassie Blue solution should have a dark blue color.

4 Notes

1. The basic theory of protein precipitation using ammonium sulfate is presented in detail by Wingfield [9]. In this reference, tables are provided to calculate the appropriate amount of ammonium sulfate to add to a particular protein solution.
2. There are many commercially available kits based on the Bradford assay or biuret test that can measure the amount of proteins. The advantages of the Bradford assay are that it is a rapid reaction and does not require a lot of handling.
3. Protect DQ-gelatin from light at all times using aluminum foil and store at −20 °C.
4. Make sure to rinse the trout with ddH_2O properly to remove all contaminants. Store the fish skin at 4 °C for up to 24 h, wrapped with parchment paper, if not used immediately. The skin obtained from a 0.5-pound trout will yield ~250 mL of pure type I collagen.

5. Culture supernatants stored at −20 °C for longer periods of time may display reduced protease activity in comparison with freshly obtained samples.
6. Use analytical grade ammonium sulfate and add small portions of solid ammonium sulfate (15–20 g) at a time to completely dissolve the salt, using gentle stirring. Avoid foaming of the solution.
7. To solubilize the precipitated material, start by adding 2 mL of buffer A and increase the amount of buffer, as needed. Avoid formation of air bubbles.
8. Cut dialysis bag according to the sample volume following manufacturer's instructions and leaving additional about 0.5 cm margin on each end. Carefully fold one end of the bag two to three times and fasten enclosing clips to secure the bag's closure. Transfer the sample carefully into the dialysis bag, repeat folding of the open end and close with another clip.
9. After exchanging buffer A for the fourth time, proceed to overnight dialysis if the resuspended sample appears clear. However, if the mixture has darker color (e.g., brown) continue to change buffer A and dialyze until the sample is translucent and light yellow. The protein may precipitate from the solution after prolonged dialysis. In that case, centrifuge the sample in 15 mL conical tubes at 5000 × *g* for 10 min. Apply the soluble fraction onto the chromatography column.
10. Prepare 10% APS solution fresh each time.
11. To prevent gel dehydration wet two to three paper towels with ddH_2O and wrap around the cassette with polymerized gel. Subsequently, place the cassette in a plastic wrap. Store at 4 °C and use within 72 h.
12. Mix ddH_2O and the 10× buffer in a 1 L graduated cylinder. Cover with Parafilm and gently mix by inverting the graduated cylinder a few times.
13. Prepare fresh each time and mix in a 50 mL conical tube.
14. Take caution when removing the gel from its casting, as it is prone to crack. Using a razor, trim the bottommost part of the gel that does not contain the protein, as well as the edges. Using a spatula carefully slide the gel from its casing into the container with 1× zymogram renaturing buffer.
15. For optimal results, incubate overnight.
16. The resulting collagen should appear dense and gelatinous. Store the collagen in a tightly sealed bottle at 4 °C and use within 6–8 months.
17. Slowly and carefully aliquot purified fish/human collagen into a 1.5 mL centrifuge tube using 1000 μL pipette tip, ensuring

that the collagen is not aspirated into the pipette. Preincubate the fish collagen at 37 °C for 5–10 min, or until the collagen appears more fluid in consistency.

Acknowledgments

This work was supported by the start-up funds from Oregon State University (OSU) as well as Faculty Development Funding to AES.

References

1. Watanabe K (2004) Collagenolytic proteases from bacteria. Appl Microbiol Biotechnol 63(5):520–526. https://doi.org/10.1007/s00253-003-1442-0
2. Harrington DJ (1996) Bacterial collagenases and collagen-degrading enzymes and their potential role in human disease. Infect Immun 64(6):1885–1891
3. Duarte AS, Correia A, Esteves AC (2014) Bacterial collagenases—a review. Crit Rev Microbiol. https://doi.org/10.3109/1040841X.2014.904270
4. Park BR, Zielke RA, Wierzbicki IH, Mitchell KC, Withey JH, Sikora AE (2015) A metalloprotease secreted by the type II secretion system links Vibrio cholerae with collagen. J Bacteriol 197(6):1051–1064. https://doi.org/10.1128/JB.02329-14
5. Halpern M, Gancz H, Broza M, Kashi Y (2003) Vibrio cholerae hemagglutinin/protease degrades chironomid egg masses. Appl Environ Microbiol 69(7):4200–4204
6. Vaitkevicius K, Rompikuntal PK, Lindmark B, Vaitkevicius R, Song T, Wai SN (2008) The metalloprotease PrtV from Vibrio cholerae. FEBS J 275(12):3167–3177. https://doi.org/10.1111/j.1742-4658.2008.06470.x
7. Vandooren J, Geurts N, Martens E, Van den Steen PE, Opdenakker G (2013) Zymography methods for visualizing hydrolytic enzymes. Nat Methods 10(3):211–220. https://doi.org/10.1038/nmeth.2371
8. Van Wart HE, Steinbrink DR (1981) A continuous spectrophotometric assay for Clostridium histolyticum collagenase. Anal Biochem 113(2):356–365
9. Wingfield P (2001) Protein precipitation using ammonium sulfate. Curr Protoc Protein Sci. Appendix 3:Appendix 3F. https://doi.org/10.1002/0471140864.psa03fs13
10. Miller EJ, Gay S (1987) The collagens: an overview and update. Methods Enzymol 144:3–41
11. Exposito JY, Valcourt U, Cluzel C, Lethias C (2010) The fibrillar collagen family. Int J Mol Sci 11(2):407–426. https://doi.org/10.3390/ijms11020407
12. Suphatharaprateep W, Cheirsilp B, Jongjareonrak A (2011) Production and properties of two collagenases from bacteria and their application for collagen extraction. New Biotechnol 28(6):649–655. https://doi.org/10.1016/j.nbt.2011.04.003
13. Zielke RA, Simmons RS, Park BR, Nonogaki M, Emerson S, Sikora AE (2014) The type II secretion pathway in Vibrio cholerae is characterized by growth phase-dependent expression of exoprotein genes and is positively regulated by sigmaE. Infect Immun 82(7):2788–2801. https://doi.org/10.1128/IAI.01292-13

Chapter 15

Proteomics of *Vibrio cholerae*

Ryszard A. Zielke

Abstract

Combining high-throughput mass spectrometry with isobaric tags for relative and absolute quantification (iTRAQ) allows for the identification and relative quantification of proteins from multiple samples. Furthermore, low-abundance proteins that are usually not detected can be enriched by using only the relevant fraction of the proteome, e.g., cytoplasmic, membrane proteins, or secreted proteins. Described here is a workflow for isolation and enrichment of secreted and membrane proteins that is compatible with mass spectrometry. Isolated proteins are reduced, alkylated, and digested with trypsin, and obtained peptides are labeled with iTRAQ reagent and separated by strong cation exchange to reduce the complexity. Finally, the peptides are separated by reverse-phase chromatography, spotted on a MALDI target plate, and analyzed by MALDI TOF-TOF.

Key words Proteomics, iTRAQ, Secreted proteins, Cell envelope, 2D-LC-MS/MS

1 Introduction

Bacterial cells interact with their surroundings through the proteins that are localized in the cell envelope. Environmental stimuli can elicit changes in the expression of many of these proteins, including membrane and secreted proteins. In the case of pathogenic bacteria, the membrane and secreted proteins allow the infection to be established, invade host cells, or evade immune responses. Many membranes and secreted proteins act as toxins, such as cholera toxin that is secreted by a type II secretion system spanning both bacterial membranes [1]. The bacterial cell envelope maintains the cell shape, and its selective permeability allows uptake of the nutrients and acts as the first line of defense against antimicrobials. Surface-associated proteins are also the target for vaccine development. Two major techniques have been used to identify membrane proteins. The first uses trypsin or other proteases to "shave" exposed protein loops and analysis of the released peptides by mass spectrometry to identify the surface-exposed proteins [2, 3]. The second method relies on enriching the samples in membrane proteins or outer membrane vesicles using differential solubility of membrane and cytoplasmic proteins and ultracentrifugation [4, 5].

Aleksandra E. Sikora (ed.), *Vibrio Cholerae: Methods and Protocols*, Methods in Molecular Biology, vol. 1839,
https://doi.org/10.1007/978-1-4939-8685-9_15, © Springer Science+Business Media, LLC, part of Springer Nature 2018

Since its introduction more than 10 years ago, iTRAQ reagent has been widely used in proteome research [6]. iTRAQ reagent tags are isobaric, composed of reporter, balancer, and peptide-reactive groups, and up to eight samples can be labeled with different isotopic variants from the tag. Samples are pooled together and analyzed by mass spectrometry at the same time. This increases the throughput of analysis because the peptides derivatized with a different tag migrate as single peak during chromatography, and their masses are indistinguishable when analyzed by mass spectrometry scan (MS1). The quantification and peptide identification is attained when the reporter group and peptide fragments are released during MS2. The fragments' ion peaks allow for peptide identification and consequently can be assigned to a single protein. The intensity of the reporter group allows for the measurement of the relative abundance of the protein in the samples. Due to the sample multiplexing and high-throughput performance, new isobaric labeling techniques have been recently reviewed [7, 8].

We used mass spectrometry combined with iTRAQ labeling to detect novel proteins secreted by type II secretion system in *Vibrio cholera* and to identify novel antigens localized to cell envelope and membrane vesicles in *Neisseria gonorrhoeae* [1, 4, 5].

2 Materials

All solution should be prepared using ultrapure water, with the sensitivity of 18 MΩ at 25 °C. All solvents used for chromatography and mass spectrometry should be of LC-MS grade.

2.1 Preparation of Proteins form Culture Supernatants

1. Beckman J-E centrifuge (Beckman Coulter).
2. 0.2 μm filter units (Thermo Fisher Scientific).
3. Complete protease inhibitor cocktail, EDTA free (Roche).
4. 1.5% sodium deoxycholate.
5. 100% (w/v) trichloroacetic acid (TCA).
6. 2D Quant Kit (GE Healthcare).
7. 100% ice-cold acetone.
8. 0.5 M triethylammonium bicarbonate buffer (TEAB) pH 8 (Sigma).
9. 2% sodium dodecyl sulfate (SDS).

2.2 Preparation of Membrane Proteins

1. Beckman J-E centrifuge (Beckman Coulter).
2. Beckman ultracentrifuge (Beckman Coulter).
3. 70 mL Beckman bottle assembly (part number: 355622) (Beckman Coulter).
4. Rotisserie.

5. 50 mL conical tubes.
6. 20× phosphate buffered saline (PBS) (Growcells).
7. cOmplete Protease Inhibitor Cocktail, EDTA free (Roche).
8. 0.1 M sodium carbonate.
9. 2D Quant Kit (GE Healthcare).
10. 100% ice-cold acetone.

2.3 iTRAQ Labeling

1. Vacuum concentrator (Eppendorf).
2. iTRAQ Reagent Multiplex Kit (contains dissolution buffer [0.5 M triethylammonium bicarbonate (TEAB)].
3. Sequencing Grade Modified Trypsin (Promega).
4. 0.1% trifluoroacetic acid.

2.4 Protein Precipitation

1. Refrigerated microcentrifuge.
2. Vortex.
3. Freezer.
4. 100% ice-cold acetone.

2.5 Sample Cleanup

1. Sep-Pak C18 columns (Waters Corporation).
2. 15 mL conical tubes.
3. 1.5 mL microcentrifuge tubes.
4. H_2O (HPLC grade).
5. Acetonitrile.
6. 0.1% TFA.
7. 0.1% TFA in 70% acetonitrile.

2.6 2D HPLC and Mass Spectrometry

1. Paradigm MG4 High-Pressure Liquid Chromatography System (Michrom Bioresources, Troy, OH) with Varian ProStar fraction collector for strong cation exchange fractionation.
2. Strong cation exchange column (2.1 × 150 mm Zorbax 300) (Agilent).
3. Nano2D LC (AB Sciex, Farmingham, MA) high-pressure liquid chromatography system coupled to Dionex Probot MALDI Spotter (Thermo Fisher Scientific).
4. NanoLC reverse-phase column (75 μm × 250 mm) packed with Magic C_{18}AQ (5-μm 100 Å resin) (Michrom Bioresources).
5. Sep-Pak C18 cartridges (1 cc/50 mg) (Waters Corporation).
6. Centrifugal vacuum concentrator.
7. Chromatography vials.

8. 4800 Proteomics Analyzer (TOF/TOF) (AB Sciex).
9. Opti-TOF LC/MALDI blank insert plate (123 × 81 mm) (AB Sciex).
10. Buffer SCX A (5 mM KH_2PO_4 pH 2.7, 30% acetonitrile).
11. Buffer SCX B (5 mM KH_2PO_4 pH 2.7, 500 mM KCl, 30% acetonitrile).
12. Buffer RP A (0.1% TFA/water).
13. Buffer RP B (0.1% TFA/90% acetonitrile).
14. 0.1% trifluoroacetic acid (TFA).
15. Matrix solution: 2 mg/ml α-cyano 4-hydroxycinnamic acid in 50:49:1 ratio of HPLC-grade water-isopropanol-acetic acid and containing 10 mM ammonium phosphate.

3 Methods

3.1 Preparation of Proteins from Culture Supernatants

1. Grow bacterial cultures with agitation at 37 °C to the late stationary phase.
2. Pellet cells from 1 L cultures (4000 × *g*, 30 min, 4 °C), and save the pellet for membrane protein extraction (*see* Subheading 3.2).
3. Filter culture supernatants through 0.22 μm filter into 1 L bottles.
4. Add one tablet of cOmplete Protease Inhibitor Cocktail, EDTA free per 1 L of culture supernatant.
5. Add 1.5% sodium deoxycholate to final concentration of 0.015%, and mix thoroughly. Place a stirring bar in each bottle.
6. To precipitate proteins, add 100% TCA to final concentration of 10%, and incubate at 4 °C for 1 h with stirring.
7. Collect precipitated proteins by centrifugation at 8000 × *g* for 15 min at 4 °C in Beckman J-E centrifuge.
8. Wash the obtained protein pellet twice with ice-cold acetone, and centrifuge at 6000 × *g* for 10 min at 4 °C in Beckman J-E centrifuge.
9. Air-dry the collected protein pellets (*see* **Note 1**).
10. Reconstitute proteins in 0.5 M TEAB 0.1% SDS (*see* **Note 2**).
11. Measure protein concentration using 2D Quant Kit (*see* **Note 3**).

3.2 Preparation of Membrane Proteins

1. Resuspend bacterial cells from Subheading 3.1, **step 1** in 20 mL of PBS, and add protease inhibitors tablet.
2. Lyse cells by passing four times through French press at 12,000 psi (*see* **Note 4**).

3. Remove cell debris by centrifugation at 10,000 × *g* for 15 min at 4 °C.
4. Measure protein concentration using 2D Quant Kit, and transfer volume containing 10 mg of proteins to a new 50 mL conical tube.
5. Add five volumes of cold 0.1 M sodium carbonate, and incubate at 4 °C for 1 h with rotation.
6. Transfer solution to 70 mL Beckman bottle assembly, and spin at 200,000 × *g* for 2 h at 4 °C in Beckman ultracentrifuge using Type 45 Ti rotor.
7. Wash protein pellet with 1× PBS and centrifuge at 200,000 × *g* for 30 min at 4 °C.
8. Remove supernatant and reconstitute proteins in 0.5 M TEAB 0.1% SDS.
9. Measure protein concentration using 2D Quant Kit.

3.3 iTRAQ Labeling

1. Transfer 100 μg of protein into microcentrifuge tube.
2. Adjust the volume to 21 μL using (0.5 M TEAB (dissolution buffer)) (*see* **Note 5**). If the volume is above 21 μL, precipitate the proteins in 90% acetone (*see* Subheading 3.4).
3. Add 2 μL of Tris(2-carboxyethyl)phosphine (TCEP), and incubate at 60 °C for 1 h.
4. Spin briefly, add 1 μL of methyl methanethiosulfonate (MMTS), and incubate at room temperature for 10 min.
5. Reconstitute trypsin in 50 mM acetic acid to a final concentration of 1 μg/μL.
6. Add 10 μL of trypsin to each sample and incubate at 37 °C overnight.
7. Allow the iTRAQ reagent to reach room temperature, and briefly spin at maximum speed to bring all the solution to the bottom of the tube.
8. Add 70 μL of absolute ethanol provided with the kit, vortex, and spin briefly.
9. Transfer the content of the iTRAQ reagent vial to a microcentrifuge containing the sample.
10. Incubate at room temperature for 1 h.
11. Add 250 μL of 0.1% TFA to each tube to quench the reaction, and incubate at room temperature for 30 min.
12. Pool all labeled samples into new microcentrifuge tube, and dry the sample in vacuum concentrator.

3.4 Protein Precipitation

1. Transfer 100 μg of proteins into a 1.5 mL microcentrifuge tube.
2. Add five volumes of ice-cold acetone and mix briefly.
3. Incubate over night at −20 °C.
4. Pellet precipitate by centrifugation at maximum speed for 15 min at 4 °C.
5. Wash protein pellet twice with ice-cold acetone, and spin at maximum speed for 10 min at 4 °C.
6. Decant acetone and air-dry pellets.

3.5 Sample Cleanup

1. Make a hole in the cap of the 15 mL conical tube and insert a C18 column.
2. Wet the column with 1 mL of acetonitrile.
3. Condition the column with 1 mL of 0.1% TFA.
4. Apply the reconstituted sample in 1 mL of 0.1% TFA.
5. Wash the column with 1 mL of 0.1% TFA.
6. Transfer the column to a new 15 mL conical tube, and elute the peptides with 0.1% TFA, 70% acetonitrile.
7. Transfer the eluate to a 1.5 mL microcentrifuge tube, and dry using the vacuum concentrator.

3.6 2D HPLC and Mass Spectrometry

1. Dissolve the sample obtained in Subheading 3.5, **step** 7 in 500 μL of Buffer SCX A.
2. Load the sample onto the equilibrated SCX column using Hamilton syringe and 500 μL loop.
3. Resolve the peptides by running 0–98% buffer SCX B gradient over 1 h with a flow rate of 200 μL/min.
4. Collect 400 μL fractions in 2 min intervals.
5. Pool fractions together to reduce the number to 12 before proceeding to RP.
6. Concentrate the samples to dryness in vacuum concentrator, reconstitute in 0.1% TFA, desalt using Sep-Pak C18 column and Subheading 3.5, and dry again.
7. Reconstitute in 10 μL of 0.1% TFA, and transfer to a chromatography vial.
8. Inject 8 μL of sample into the nanoHPLC, and run samples using 90 min gradient from 7 to 35% ACN in 0.1% TFA at a flow rate of 500 mL/min using a reversed-phase column (75 μm × 250 mm) packed with Magic C18AQ directly mounted to the Dionex Probot MALDI Spotter.

9. Mix the column effluent using a micro-Tee with matrix delivered at 1 μL/min, and spot fractions at 10 s intervals onto a stainless steel blank MADLI target plate.
10. Spot peptide standards onto a MALDI target plate.
11. Place the MALDI target plate in the plate holder inside the 4800 Proteomics Analyzer.
12. Load and align the plate.
13. Adjust the laser power high enough to produce high-quality mass spectra but low enough to still provide the best possible resolution.
14. Acquire the MS mass spectra.
15. Acquire MS/MS spectra.
16. Remove the plate from the instrument.

4 Notes

1. Dried protein pellets can be stored at −80 °C indefinitely.
2. Buffers containing amine groups, e.g., Tris buffers, ammonium salts, or 2-amino-2methyl-1-propanol (AMP), should be avoided as they react with the *N*-Hydroxysuccinimide group present in the iTRAQ labeling.
3. Alternatively, other kits compatible with denaturating agents can be used (e.g., Pierce BCA Protein Assay Kit or Bio-Rad DC Protein Assay Kit).
4. The French press cell should be placed at 4 °C overnight to avoid sample overheating. If a French press is not available, other methods of cell lysis can be used such as sonication or the freeze-thaw method.
5. The iTRAQ Reagent Kit provides all of the necessary solutions: dissolution buffer (0.5 M TEAB), denaturant (2% SDS), reducing agent (50 mM Tris(2-carboxyethyl)phosphine), cysteine-blocking reagent (200 mM methyl methanethiosulfonate in isopropanol), and ethanol.

References

1. Sikora AE, Zielke RA, Lawrence DA, Andrews PC, Sandkvist M (2011) Proteomic analysis of the Vibrio cholerae type II secretome reveals new proteins, including three related serine proteases. J Biol Chem 286:16555–16566. https://doi.org/10.1074/jbc.M110.211078
2. Doro F et al (2009) Surfome analysis as a fast track to vaccine discovery: identification of a novel protective antigen for Group B Streptococcus hypervirulent strain COH1. Mol Cell Proteomics 8:1728–1737. https://doi.org/10.1074/mcp.M800486-MCP200

3. Rodriguez-Ortega MJ et al (2006) Characterization and identification of vaccine candidate proteins through analysis of the group A Streptococcus surface proteome. Nat Biotechnol 24:191–197. https://doi.org/10.1038/nbt1179
4. Zielke RA et al (2016) Proteomics-driven antigen discovery for development of vaccines against gonorrhea. Mol Cell Proteomics. https://doi.org/10.1074/mcp.M116.058800
5. Zielke RA, Wierzbicki IH, Weber JV, Gafken PR, Sikora AE (2014) Quantitative proteomics of the Neisseria gonorrhoeae cell envelope and membrane vesicles for the discovery of potential therapeutic targets. Mol Cell Proteomics 13:1299–1317. https://doi.org/10.1074/mcp.M113.029538
6. Ross PL et al (2004) Multiplexed protein quantitation in Saccharomyces cerevisiae using amine-reactive isobaric tagging reagents. Mol Cell Proteomics 3:1154–1169. https://doi.org/10.1074/mcp.M400129-MCP200
7. Chahrour O, Cobice D, Malone J (2015) Stable isotope labelling methods in mass spectrometry-based quantitative proteomics. J Pharm Biomed Anal 113:2–20. https://doi.org/10.1016/j.jpba.2015.04.013
8. Rauniyar N, Yates JR 3rd (2014) Isobaric labeling-based relative quantification in shotgun proteomics. J Proteome Res 13:5293–5309. https://doi.org/10.1021/pr500880b

Index

A

Accelerated evolution 53
Affinity chromatography 157

B

Biochemical assays 11
Biotypes 2, 11–26, 50
Bradford 118–120, 123–127, 131, 167

C

cDNA synthesis 30, 31, 34, 35
Cell envelope 136, 171, 172
Cholera 1–7, 11–26, 29–51, 53–63, 65–75, 77–95, 97–101, 103–115, 117, 119, 130–132, 135–144, 147–151, 153–169, 171–177
Chromatin immunoprecipitation (ChIP) 65–75
Classical 2, 4–7, 11–17, 23, 50
Collagenase 153–169
Collagenolytic proteases 153–169
Colonization 2, 12, 77–95, 98, 100, 101, 147–151
Combinatorial mutagenesis 53
Competitive index 148, 150
Confocal microscopy 86, 92
Cotransformation 53–63

D

Density gradient purification 118, 123–124
Diarrhea 1, 2, 11, 45, 98, 105, 135
2D-LC-MS/MS 171
DNA-binding proteins 65, 66, 69, 72, 73
DQ-gelatin 154–156, 159, 165, 167
Drosophila melanogaster 83–84

E

El Tor 2–7, 11–17, 23
El Tor variants 2, 12–17

F

Fruit fly 77
Furanacryloyl-L-leucylglycyl-L-prolyl-L-alanine (FALGPA) 154, 156–157, 160, 166

G

Growth 3–7, 13, 23, 24, 26, 46, 61–63, 73, 78, 86, 87, 93–95, 105, 137, 139–142, 154, 155, 157, 158, 162, 166

H

High-throughput screening 137–138, 140–142
House fly 79, 81–83, 89–92

I

Illumina 30, 31, 35, 36, 106, 107, 112–114
Immunoblot 118, 119, 121–122, 124, 127–128, 133
Immunoprecipitation 65–75
Infant rabbit 103–115
Infectious disease transmission 97
Isobaric tags for relative and absolute quantification (iTRAQ) 172, 173, 175, 177

L

Laboratory 1–7, 78, 79, 94, 122, 137, 141–143
Lipid extraction 118, 119, 122–123, 128–130
Liquid chromatography-mass spectrometry (LC-MS) 45–51, 118, 129–130, 137, 139, 140, 142–144, 172

M

Maintenance 1–7, 89–90, 92, 104
Mariner transposon 39, 42
Metalloprotease 153, 154
Mouse model 105, 147–151
mRNA enrichment 30, 32–34
Multiplex genome editing 53–63
Musca domestica 77, 79, 89–91

N

NanoSight 118, 120–121
Natural products 135–144
Natural transformation 53–63, 107
Nested PCR 42
Next-generation sequencing 29, 66

Aleksandra E. Sikora (ed.), *Vibrio Cholerae: Methods and Protocols*, Methods in Molecular Biology, vol. 1839, https://doi.org/10.1007/978-1-4939-8685-9, © Springer Science+Business Media, LLC, part of Springer Nature 2018

O

Outer membrane vesicles (OMVs) 117–133, 171

P

Phenyl-bonded phase 45
Proteomics 171–177
Purpald 118, 120, 125–126, 132

Q

Q-TOF mass spectrometer 47, 50
Quantification 37, 115, 172

R

Random mutagenesis 39–44
Random-primed sequencing 40–43
RNA-seq 29–38

S

Secreted proteins 136, 171, 172
Secretion 117, 135, 136, 157, 171, 172
Serine proteases 137, 138
Sodium dodecyl sulfate-polyacrylamide gel electrophoresis (SDS-PAGE) 118, 121, 122, 124, 126–127, 157–159, 163, 164, 167
Survival 80, 84–87, 135

T

Transcription 29, 65, 72, 75
Transcriptomic analysis 29
Transposon mutagenesis 39–44
Transposon sequencing (Tn-seq) 103–115
TrayCell™ 118, 120, 124, 125, 131
Type I collagen 153–169
Type II secretion system 171, 172

U

Ultracentrifugation 123, 171
Untargeted metabolomics 46, 47, 50

V

Vibrio cholera 1–7, 11–26, 29–51, 53–63, 65–75, 77–95, 97–101, 103–115, 135–144, 147–151, 153–169, 171–177
Virulence 2–4, 6–7, 12, 13, 40, 45, 51, 77, 78, 86, 94, 117, 135, 136, 147, 148
Virulence inducing 2–4, 6–7

Z

Zebrafish 97–101
Zymography 154, 155, 159, 164–165

MIX
Papier aus verantwortungsvollen Quellen
Paper from responsible sources
FSC® C105338

If you have any concerns about our products,
you can contact us on
ProductSafety@springernature.com

In case Publisher is established outside the EU,
the EU authorized representative is:
Springer Nature Customer Service Center GmbH
Europaplatz 3, 69115 Heidelberg, Germany

Printed by Libri Plureos GmbH
in Hamburg, Germany